Medical Malpractice
Understanding the Law, Managing the Risk

Medical Malpractice
Understanding the Law, Managing the Risk

S. Y. Tan, MD, JD
University of Hawaii, USA

NEW JERSEY · LONDON · SINGAPORE · BEIJING · SHANGHAI · HONG KONG · TAIPEI · CHENNAI

Published by

World Scientific Publishing Co. Pte. Ltd.
5 Toh Tuck Link, Singapore 596224
USA office: 27 Warren Street, Suite 401-402, Hackensack, NJ 07601
UK office: 57 Shelton Street, Covent Garden, London WC2H 9HE

Library of Congress Cataloging-in-Publication Data
Tan, Siang Yong.
  Medical malpractice : understanding the law, managing the risk / S.Y. Tan
    p. cm.
  Includes bibliographical references and index.
  ISBN-13 978-981-256-668-3
  ISBN-10 981-256-668-6
  1. Medical personnel--Malpractice--United States. I. Title.

KF2905.3.Z9T36 2006
344.7304'11--dc22

2006040353

British Library Cataloguing-in-Publication Data
A catalogue record for this book is available from the British Library.

Copyright © 2006 by World Scientific Publishing Co. Pte. Ltd.

All rights reserved. This book, or parts thereof, may not be reproduced in any form or by any means, electronic or mechanical, including photocopying, recording or any information storage and retrieval system now known or to be invented, without written permission from the Publisher.

For photocopying of material in this volume, please pay a copying fee through the Copyright Clearance Center, Inc., 222 Rosewood Drive, Danvers, MA 01923, USA. In this case permission to photocopy is not required from the publisher.

Typeset by Stallion Press
Email: enquiries@stallionpress.com

Printed in Singapore

Dedication

To Babes, for devotion and encouragement

To Sister Beatrice Tom, CEO, St. Francis Healthcare System,

for steadfast trust and support

Acknowledgements

I thank first and foremost the medical students who assisted me in the research and preparation of parts of this book. They inspired and kept the fire burning, and inspired more than they realize. These students, now doctors in their own right, are Reina Ahern, Jason Fleming, Sheri Meslinsky, Stefanie Park, Diane Pratt and Laura Rogers. Their specific works are acknowledged within the text but I would be remiss not to mention them here as well.

The following colleagues were reviewers of the manuscript, and they identified errors of fact or law, offered helpful suggestions on style, and in general greatly improved the book. I, of course, bear full responsibility for any residual errors or oversight. These colleagues include Tina Bails, Hazel Beh, Edmund Burke, Louise Chiu, Elizabeth Clark-Palomo, Jeffrey Crabtree, Collin Dang, George Druger, Vanessa Duplechain, Richard Frankel, Rick Fried, Erin Fuse-Brown, Ellen Godbey-Carson, Joy Graham, Gary Hageman, Robert Harris, Kelvin Kaneshiro, Marshall Kapp, David Karp, Richard Kasuya, Ronald Libkuman, Allen Llyod, Anderson Meyer, John Nishimoto, Joe Peppin, Jim Pietsch, Ken Robbins, Rene Shimabukuro, Myron Shirasu, Lois Snyder, Leilani Tan-Ching, Douglas Thom, Alan Tice, the Honorable Judge Michael Town, and Karen Uyeda.

I am grateful to my able assistant Jan Miyamoto for proof-reading the manuscript in its entirety, and to Debbie Miyajima for secretarial help.

Finally, special thanks to Ms. Lim Sook Cheng and the staff at World Scientific Publishing for guidance, timeliness and unmatched competence.

Contents

Table of Cases		xi
About the Author		xix
Introduction		xxi
SECTION I:	**THE LAW OF MALPRACTICE AND INFORMED CONSENT**	1
Chapter 1	Condition Critical, Prognosis Guarded	3
Chapter 2	What Is and Isn't Malpractice	21
Chapter 3	Legal Duty	28
Chapter 4	Standard of Care	37
Chapter 5	Causation	48
Chapter 6	Damages	56
Chapter 7	Expert Testimony	63
Chapter 8	Affirmative Defenses	73
Chapter 9	Physician Countersuits	79
Chapter 10	Informed Consent	88
Chapter 11	Patients Who Drive: A New Worry for Doctors	104
Chapter 12	End-of-Life Issues	111

Chapter 13	Complementary and Alternative Medicine	131
Chapter 14	Products Liability	139
Chapter 15	Institutional and Managed Care Liability	148
Chapter 16	Medical Trainees	157

SECTION II: RISK MANAGEMENT — 169

Chapter 17	Medical Records and Confidentiality	171
Chapter 18	Cyber-Medicine	182
Chapter 19	What Do Malpractice Lawsuits Look Like	187
Chapter 20	How to Avoid Becoming a Horror Story	198
Chapter 21	Medical Errors and Their Disclosure	217
Chapter 22	Understanding Malpractice Insurance	236
Chapter 23	What to Expect Now That You've Been Sued	246

SECTION III: REFORMING THE SYSTEM — 259

Chapter 24	Medical and Legal Reforms	261

SECTION IV: MULTIPLE CHOICE QUESTIONS AND ANSWERS — 281

SECTION V: GLOSSARY OF LEGAL TERMS — 319

INDEX — 335

Table of Cases

	Page
Ackerman v. Lagano, 412 A.2d 1954 (N.J. 1979).	82
Advincula v. United Blood Servicers, 678 N.E.2d 1009 (Ill. 1996).	46
Aetna Health Inc., v. Davila, and *Cigna Healthcare of Texas v. Calad*, 124 S. Ct. 2488 (2004).	155
Allinson v. General Council Medical Education and Registration, 1 Q.B. 750 (1894).	24
Arato v. Avedon, 858 P.2d 598 (Cal. 1993).	92
Austin v. American Assn. of Neurological Surgeons, 253 F.3d 967 (7th Cir., Ill. 2001).	66
Barber v. Superior Court, 147 Cal.App.3d 1006 (1985).	25, 128
Barrett v. Harris, 86 P.3d 954 (Ariz. 2004).	50
Benison v. Silverman, 599 N.E.2d 1101 (Ill. 1992).	64
Bergman v. Eden Medical Center, No. H205732-1 (Sup. Ct. Alameda Co., Cal. 2001).	122, 196
Berlin v. Nathan, 64 Ill.App.3d 940 (1st Dist. 1978).	81
Bernard v. Char, 903 P.2d 667 (Haw. 1995).	102
Bird v. Saenz, 103 Cal.Rptr.2d 131 (Cal. App. 2001).	52
Blanchard v. Kellum, 975 S.W.2d 522 (Tenn. 1998).	89
Bobbitt v. Chow, 65 P.3d 182 (Haw. 2003).	70
Bolam v. Friern Hospital Management Committee, 1 WLR 582 (1957).	23
Boody v. United States, 706 F.Supp.1458 (D.Kan. 1989).	52
Boyd v. Albert Einstein Medical Center, 547 A.2d 1229 (Penn. 1988).	153
Boyle v. Revici, 961 F.2d 1060 (2nd Cir., N.Y. 1992).	135

(*Continued*)

	Page
Bradshaw v. Daniel, 854 S.W.2d 865 (Tenn. 1993).	33
Brown v. Bower, No. J86-0759 (B), SD. Miss. Dec. 21, 1987.	35
Brown v. Flowe, 496 S.E.2d 830 (N.C. App. 1998).	157
Bryan v. Burt, 486 S.E.2d 536 (Va. 1997).	24
Bull v. McCuskey, 96 Nev. 706 (1980).	84
Butterfield v. Forrester, 103 ER 926 (1809).	75
Byrne v. Boadle, 2 H. & C. 722, 159 Eng. Rep. 299 (Court of the Exchequer 1863).	43
Cangemi v. Cone, 774 A.2d 1262 (Pa. Super. Ct. 2001).	43
Canterbury v. Spence, 464 F.2d 772 (D.C. Cir. 1972).	96, 100
Carlisle v. Carnival Corporation, et al., 864 So.2d 1 (Fl. 2003).	151
Carr v. Strode, 79 Hawaii 475 (1995).	101
Cassidy v. Ministry of Health, 2 KB 343 (1951).	44
Chaffin v. Nicosia, 310 N.E.2d 867 (Ind. 1974).	74
Charrell v. Gonzales, 673 N.Y.S. 2d 685 (1998).	134
Cobbs v. Grant, 502 P.2d 1 (Cal. 1972).	100
Colorado Chiropractic Council v. Porter Memorial Hospital, 650 F. Supp. 231 (Co. 1986).	85
Cominsky v. Donovan, 846 A.2d 1256 (Super. Ct. Penn. 2004).	63
Craft v. Peebles, 893 P.2d 138 (Haw. 1995).	41
Crescenzo v. Crane, 796 A.2d 283 (N.J. Super. 2002).	180
Cruzan v. Director, Missouri Department of Health, 110 S. Ct. 2841 (1990).	114
Cummins v. Rosa, 846 A.2d 148 (Super. Ct. Penn. 2004).	68
Darling v. Charleston Community Memorial Hospital, 211 N.E.2d 253 (Ill. 1965).	149
Daubert v. Merrell Dow Pharmaceurticals, Inc., 113 S. Ct. 2786 (1993).	69
Dempsey v. Phleps, 700 So.2d 1340 (Ala. 1997).	59
Desiderio v. Ochs, 741 N.Y.S.2d 865 (2002); *affirmed* 791 N.E.2d 941 (N.Y. 2003).	16
Doe v. Sullivan, 938 F.2d 1370 (D.C. Cir. 1991).	97
Domingo v. T.K., 289 F.3d 600 (9th Cir., Haw. 2002).	70
Dotson v. Hammerman, 932 S.W.2d 880 (Mo. App. 1996).	23
Drago v. Buonagurio, 391 N.Y.S.2d 61 (Sup. Ct. Schenectady Co. 1977).	84

(Continued)

	Page
Dulieu v. White, 2 K.B. 669 (1901).	53
Dutt v. Kremp, 111 Nev. 567 (1995).	81
Estate of Behringer v. The Medical Center at Princeton, 192 A.2d 1251 (N.J. Super. 1991).	92
Estate of Henry James v. Hillhaven Corp., Super Ct. Div. 89 CVS 64 (Hertford Cty., N. C. Nov. 20, 1990).	123
Evanston Hospital v. Crane, 627 N.E.2d 29 (Ill. App. 1993).	43
Felice v. Valleylab, Inc., 520 So.2d 920 (La. Ct. App. 1988).	165
Fellin v. Sahgal, 745 N.Y.S.2d 565 (2002); dismissed (*JNOV*) 798 N.Y.S.2d 344 (2004).	16
Freese v. Lemmon, 210 N.W.2d 576 (Iowa 1973).	106
Gentzler v. Atlee, 443 Pa. Super. 128 (1995).	81
Giallanza v. Sands, 316 So.2d 77 (Fl. App. 4 Dist. 1975).	29
Gilgunn v. Massachusetts General Hospital, Super. Ct. Civ. Action No. 92-4820, Suffolk Co., Mass., verdict, 21 April, 1995.	120
Glenn v. Plante, 269 Wis.2d 575 (2004).	65
Gonzales v. Nork, Suit No. 22856, Super. Ct. Cal. (1973).	149
Gonzales v. NYS DOH, 232 A.D.2d 886 (N.Y. 1996).	136
Gordon v. Glass, 785 A.2d 1220 (Conn. App. 2001).	43
Grant v. American Nat. Red Cross, 745 A.2d 316 (D.C.App. 2000).	52
Greenman v. Yuba Power Products Inc., 377 P.2d 897 (Cal. 1963).	140
Gunapathy Muniandy v. Dr. James Khoo et al., Civil Appeal No. 600094 of 2001, Court of Appeal, Singapore.	6
H.L. v. Matheson, 450 U.S. 398 (1981).	94
Hadley v. Terwilleger, 873 So.2d 378 (Fl. 2004).	51
Hale v. Venuto, 137 Cal.App.3d 910 (Cal.App. 1982).	44
Hamilton v. Bares, 678 N.W.2d 74 (Neb. 2004).	23
Harland v. State of California, 75 Cal.App.3d 475 (1977).	105
Helling v. Carey, 519 P.2d 981 (Wash. 1974).	45
Henderson v. Heyer-Schulte Corp. of Santa Barbara, 600 S.W.2d 844 (Tex Civ. App. 1980).	40
Herskovits v. Group Health Co-op. of Puget Sound, 664 P.2d 474 (Wash. 1983).	52
Hidding v. Williams, 578 So.2d 1192 (La. App. 1991).	92
Hines v. St. Joseph's Hospital, 527 P.2d 1075 (N. Mex. 1974).	143

(Continued)

	Page
Hirahara v. Tanaka, 959 P.2d 830 (Haw. 1998).	24
Holloway v. Texas Medical Association, 757 S.W.2d 810 (Tex. 1988).	83
In re Baby K, 16 F.3d 590 (4th Cir., Va. 1994).	120
In re Baby K, 832 F. Supp. 1022 (4th Cir., Va. 1993).	120
In re Helga Wanglie, Fourth Judicial District (Dist. Ct., Probate Ct. Div.), PX-91-283. Minnesota, Hennepin County, 1991.	120
In the matter of Karen Quinlan, 355 A.2d 647, (N.J. 1976).	112, 129
Jackson v. Taylor, 912 F.2d 795 (5th Cir., Tex. 1990).	59
Jistarri v. Nappi, 549 A.2d 210 (Pa. Super. 1988).	159
Johnson v. Kokemoor, 545 N.W.2d 495 (Wis. 1996).	91
Keebler v. Winfield Carraway Hospital, 531 So.2d 841 (Ala. 1988).	37
Kennedy v. Parrott, 90 S.E.2d 754 (N.C. 1956).	95
Keomaka v. Zakaib, 811 P.2d 478 (Haw. 1991).	93, 97
Kovacs v. Pritchard, Santa Clara Cty. Sup. Ct No. CV791479, April 8, 2004.	231
Kraus v. Cleveland Clinic, 442 F.Supp. 310 (Ohio 1977), referring to R.C. Ohio §2305.11(A).	74
Kumho Tire v. Carmichael, 526 U.S. 137 (1999).	69
Landeros v. Flood, 551 P.2d 389 (Cal. 1976).	45
Larsen v. State Sav. & Loan Ass'n, 640 P.2d 286 (Haw. 1982)	70
Leach v. Drummond Medical Group, Inc., 144 Cal. App.3d 362 (1983).	34
Lection v. Dyll, 65 S.W.3d 696 (Tex. App. Dallas 2001).	29
Lejeunee v. Rayne Branch Hospital, 556 So.2d 559 (La. 1990).	38
Lester v. Hall, 126 N.M. 405 (1998).	105
Leyson v. Steuermann, 705 P.2d 37 (Haw. 1985).	100
Lomeo v. Davis, 53 Pa. D. & C.4th 49 (Pa. Com. Pl. Jul 24, 2001).	75
Longman v. Jasiek, 414 N.E.2d 520 (Ill. App. 1980).	34
Lownsbury v. VanBuren, 762 N.E.2d 354 (Ohio 2001).	163
Martin v. Herzog, 228 N.Y. 164 (1920).	45
Maxwell v. Cole, 482 N.Y.S.2d 1000 (N.Y. Sup. Ct. 1984).	165
McCarthy v. Boston City Hospital, 266 N.E. 2d 292 (Mass. 1971).	45

Table of Cases

(Continued)

	Page
McCullough v. Hutzel Hospital, 276 N.W.2d 569 (Mich. App. 1979).	162
McKenzie v. Hawaii Permanente Medical Group, Inc., 47 P.3d 1209 (Haw. 2002).	33, 104
Mercer v. Vanderbilt University, Inc., 134 S.W.3d 121 (Tenn. 2004).	76
Miller v. Rosenberg, 749 N.E.2d 946 (Ill. 2001).	83
Miskew v. Hess, 21 Kan.App.2d 927 (1996).	82
Mitchell v. Gonzales, 819 P.2d 872 (Cal. 1991).	48
Miyamoto v. Lum and Lazo, 84 P.3d 509 (Haw. 2004).	133
Mohr v. Williams, 104 N.W. 12 (Minn. 1905).	95
Moore v. Baker, 989 F.2d 1129 (11th Cir., Ga. 1993).	135
Moore v. The Regents of the University of California, 793 P.2d 479 (Cal. 1990).	92
Morowitz v. Marvel, 423 A.2d 196 (D.C. 1980).	82
Mulder v. Parke Davis & Company, 181 N.W.2d 882 (Minn. 1970).	40
Myers v. Quesenberry, 193 Cal.Rptr. 733 (1983).	106
Nisenholtz v. Mount Sinai Hospital, 483 N.Y.S.2d 568 (1984).	93
Nishi v. Hartwell, 473 P.2d 116 (Haw. 1970).	96, 99
O'Neal v. Hammer, 953 P.2d 561 (Haw. 1998).	92
O'Neil v. Montefiore Hospital, 11 A.D.2d 132 (N.Y.A.D. 1 Dept. 1960).	29
Oliver v. Brock, 342 So.2d 1 (Ala. 1976).	29
Palsgraf v. Long Island R. R. Co., 162 N.E. 99 (Ct.App.N.Y. 1928).	28, 50
Parmelee v. Kline, 579 So.2d 1008 (La. App. 1991).	163
Payton v. Weaver, 131 Cal.App.3d 38 (1982).	35
Pederson v. Dumouchel, 431 P.2d 973 (Wash. 1967).	46
Pegram v. Herdrich, 120 S. Ct. 2143 (2000).	154
People v. Einaugler, 208 A.D.2d 946 (N.Y. 1994).	25
People v. Montemarano, Indictment No. 37707, Nassau County Court, N.Y. (1974).	128
Perkins v. Howard, 232 Cal.App.3d 708 (1991).	30
Perlmutter v. Beth David Hospital, 123 N.E.2d 792 (N.Y. 1954).	143

(Continued)

	Page
Planned Parenthood of Central Mo. v. Danforth, 428 U.S. 52 (1976).	94
Pollard v. Goldsmith, 572 P.2d 1201 (Ariz. App. 1977).	42
Pratt v. Stein, 444 A.2d 674 (Pa. Super. 1980).	158
Prosise v. Foster, 544 S.E.2d 331 (Va. 2001).	163
Purtill v. Hess, 489 N.E.2d 867 (Ill. 1986).	37
R v. Adomako, 2 All ER 79 (1994).	26
Raine v. Drasin, 621 S.W.2d 895 (Ky. 1981).	82
Rainer v. Grossman, 31 Cal.App.3d 539 (Cal. App. 2 Dist. 1973).	164
Raleigh Fitkin-Paul Morgan Memorial Hospital v. Anderson, 201 A.2d 537 (N.J. 1964).	89
Rex v. Bateman, 19 Cr. App. Rep. 8 (1925).	26
Richardson v. Denneen, 82 N.Y.S.2d 623 (N.Y. Super. 1947).	163
Rockwell v. Stone, 173 A.2d 48 (Pa. Super. 1961).	162
Roskin v. Rosow, # 301356, San Mateo Cty Super. Ct., Cal. April 1987.	49
Rush v. Akron General Hospital, 171 N.E.2d 378 (Ohio Ct. App. 1987).	158
Salazar v. Ehmann, 505 P.2d 387 (Co. 1972).	133
Salgo v. Leland Stanford Jr. Univ. Board of Trustees, 317 P.2d 170 (Cal. 1957).	41
Schloendorff v. Society of New York Hospital, 101 N.E. 92 (N.Y. 1914).	89
Schneider v. Revici, 817 F.2d 987 (2nd Cir., N.Y. 1987).	77, 135
Scott v. Wilson, 396 S.W.2d 532 (Tex. Civ. App. 1965).	91
Seitz v. Akron Clinic, 557 N.E.2d 1216 (Ohio 1990).	64
Selig v. Pfizer, 713 N.Y.S.2d 898 (2000).	68
Sheldon Appel Co. v. Albert & Oliker, 47 Cal.3d 863 (1989).	81
Shepard v. Redford Community Hospital, 390 N.W.2d 239 (Mich. App. 1986).	32
Shivers v. Good Shepard Hospital, 427 S.W.2d 104 (Tex. Civ. App. 1968).	144
Siebe v. University of Cincinnati, 766 N.E.2d 1070 (Ohio Ct.Cl. 2001).	165

(Continued)

	Page
Siggers v. Barlow, 906 F.2d 241 (6th Cir., Ky. 1990).	51
Simonsen v. Swenson, 177 N.W. 831 (Neb. 1920).	180
Simpson v. Davis, 549 P.2d 950 (Kan. 1976).	39
Snyder v. Am. Ass'n. of Blood Banks, 676 A.2d 1036 (N.J. 1996).	46, 144
Spencer v. Burglass, 337 So.2d 596 (La. App. 1976).	82
Stahlin v, Hilton Hotels Corp., 484 F.2d 580 (7th Cir., Ill. 1973).	45
Stopka v. Lesser, 82 Ill. App.3d 323 (1st Dist. 1980).	82
Swidryk v. St. Michael's Medical Center, 493 A.2d 641 (N.J. Super. 1985).	164
Sylvia v. Gobeille, 220 A.2d 222 (R.I. 1966).	32
Tarasoff v. Regents of University of California, 551 P.2d 334 (Cal. 1976).	33, 179
Teresa Vasquez et al. v. Ramachandra Kolluru, Ector County (TX) District Court Case No. A-103,042 (1999).	174
United Blood Services v. Quintana, 827 P.2d 509 (Co. 1992)	46
United States v. Frye, 293 F. 1013 (D.C. Cir. 1923).	67
Vacco v. Quill, 117 S. Ct. 2293 (1997).	123
Walker v. Skiwski, 529 So.2d 184 (Miss. 1988).	44
Wallman v. Kelley, 976 P.2d 330 (Co. 1998).	133
Warren v. Schecter, 67 Cal.Rptr.2d 573 (Cal. 1997).	59, 91
Washington v. Glucksberg, 117 S. Ct. 2258 (1997).	125
Weil v. Seltzer, 873 F.2d 1453 (D.C. Cir. 1989).	76
Wendland v. Wendland, 28 P.3d 151 (Cal. 2001).	115
Wengel v. Herfert, 473 N.W.2d 741 (Mi. 1991).	133
Wickline v. State, 192 Cal.App.3d 1630 (1986).	153
Williams v. Coombs, 224 Cal.Rptr. 865 (Cal. App. 1986).	80
Wilschinsky v. Medina, 108 N.M. 511 (1989).	105
Wolfe v. Arroyo, 543 S.W.2d 11 (Tex. Ct. App. 1976).	84
Ybarra v. Spangard, 154 P.2d 687 (Cal. 1944).	44, 50
Yeats v. Harms, 393 P.2d 982 (Kan. 1964).	91

About the Author

Dr. S. Y. Tan studied medicine at McGill and Yale, and is Board certified in Internal Medicine and in Endocrinology. He is also a licensed attorney, having graduated from the University of Hawaii William S. Richardson School of Law, where he served on Law Review. Both a Professor of Medicine and an Adjunct Professor of Law at the University of Hawaii, he directs The St. Francis International Center for Healthcare Ethics and is the current Chief of Medicine at St. Francis Medical Center.

S. Y. has written over 100 articles, and lectures widely on medical, medico-legal and bioethical topics. He has been qualified as a medical expert at trial, represented clients in malpractice and other health law issues, and served on the Hawaii Medical Claims Conciliation Panel. He lives with his wife, Alexandria, in Honolulu and has two children, Leilani and Quincy.

Introduction

Medical Malpractice: Understanding the Law, Managing the Risk is about the law of medical malpractice and how to prevent a malpractice lawsuit. It grew out of an earlier book covering medical negligence in the Republic of Singapore. The book's primary goal is to provide a clear and simple explanation of the American law of medical malpractice, informed consent and risk management. It is written with the clinician in mind, so I have kept it legally uncomplicated without being overly simplistic. The reader should not construe the contents as medical or legal advice, and is urged to consult licensed practitioners for any and all questions of a more specific or personal nature.

The book is divided into five sections. Section I covers the law of malpractice and informed consent and is necessarily the most legalistic and technical part of the book. Section II covers risk management with chapters on confidentiality, communication and risk management tips. In many ways, it is the most important section because it speaks to the prevention of a lawsuit. Section III is a single chapter on reforming the system, and discusses both medical and legal proposals. The subject of tort reforms is covered in this chapter. A review section consisting of 35 multiple choice questions and answers constitutes Section IV. This feature is meant to provide a quick review of key topics. The book ends with Section V which features a glossary of legal terms.

Medical Malpractice is as much about medicine as it is about law. It is above all, about patients. I have written it with the fervent belief that with better education, there will emerge a better appreciation of the expectations of the patient — often unmet, and the standards of the legal system — often misunderstood. Fewer lawsuits and improved patient care will hopefully follow.

S. Y. Tan, MD, JD
Hawaii, 2006

SECTION I

THE LAW OF MALPRACTICE AND INFORMED CONSENT

1

Condition Critical, Prognosis Guarded

Being sued is a stressful experience. Indeed, the mere threat of a lawsuit is enough to generate high anxiety, yet all physicians expect to be sued for medical malpractice at least once during their career. Soaring malpractice insurance premiums in the 1970s created a crisis, with many doctors modifying their practice patterns or taking early retirement. Currently, another crisis is upon us. Rates have taken a quantum leap in the past couple of years, and doctor protests and walkouts signal the onset of yet another round of blame-the-lawyer, blame-the-insurer, blame-the-doctor.

After a decade of relative stability, malpractice premiums have recently increased by an average of nearly 15% nationwide, with some states seeing a disproportionate increase. West Virginia, for example, saw its average premium for all physicians rise by 35% in the year 2000. The situation is even more acute for its internists, who have to pay 50% more. In 2001, a 13-physician multi-specialty internal medicine group in Morgantown, West Virginia saw its annual premium jump from $167,000 to $288,000. If the group had stayed with its former insurer, the price tag would have been even higher at $422,000.[1]

Doctors in metropolitan areas tend to pay the highest premiums. Miami is the hardest hit with internists paying as high as $65,000 a year; general surgeons, $227,000; and obstetricians, $250,000. These are annual premiums for the standard $1 million/$3 million claims-made policy. Detroit, Chicago, Houston and Philadelphia are only slightly better. Tables 1 and 2, adapted

---

[1] ACP-ASIM Observer, April 2001, p. 1.

Table 1: States with Highest Malpractice Rates, 2003

	Internists	Surgeons	ObGyn
Florida (Miami area)	$65,697	$226,542	$249,196
Michigan (Detroit area)	$50,063	$145,165	$154,165
Illinois (Chicago area)	$41,238	$98,319	$147,023
Texas (Houston area)	$34,346	$109,668	—
Pennsylvania (Philadelphia area)	$29,667	$131,348	$152,730
Nevada (Las Vegas area)	—	—	$141,704

Table 2: States with Lowest Malpractice Rates, 2003

	Internists	Surgeons	ObGyn
Nebraska	$2,786	$9,474	$16,194
Minnesota	$3,375	$8,717	$18,307
S. Dakota	$3,697	$9,597	$14,662
Idaho	$3,770	$14,514	$19,320
Kansas	$4,123	—	—
Iowa	—	$14,386	—
Oklahoma	—	—	$22,454

from *Medical Economics*, list the states with the highest and lowest rates for three specialties.[2] These data are taken from surveys published by Medical Liability Monitor, a trade newsletter.[3]

Not all states are similarly affected. Surprisingly, some rural states such as Mississippi and Arkansas have seen an almost doubling of premiums, and within a given state, premiums can vary greatly depending on risk experience.

---

[2] *Medical Economics*, January 9, 2004.
[3] *Medical Liability Monitor*, 2003 Rate Survey. Copies of full report are available from MLM at 312-944-7900, by fax at 312-944-8845, or by email at cptmdgolin@aol.com.

Time magazine published in its June 9, 2003 issue the average annual premiums in five different medical specialties[4]:

Neurosurgeon	$71,200
ObGyn	$56,546
Emergency physician	$53,500
Orthopedic surgeon	$38,000
General surgeon	$36,354

The article also identified cities with the highest premiums. For example, the premium for neurosurgeons in Chicago was $283,000 and in Philadelphia, $267,000. In Miami, the general surgeon pays $174,268, and in Detroit $107,139.

According to the American Medical Association (AMA), one in six doctors face a malpractice claim each year, but this figure has been challenged. The AMA recently named 19 states as being in crisis; these are Arkansas, Connecticut, Florida, Georgia, Illinois, Kentucky, Mississippi, Missouri, Nevada, New Jersey, New York, North Carolina, Ohio, Oregon, Pennsylvania, Texas, Washington, West Virginia, and Wyoming. On June 14, 2004, the AMA added Massachusetts as the 20th crisis state.[5]

Malpractice damages are widely reported to reach the multi-million dollar level. While this is true in some cases of serious injuries such as neonatal brain injuries, the typical award for other medical injuries is substantially less. However, the severity of both awards and settlements has climbed steadily since 1993, up some 60%. In 1999, one in four malpractice verdicts reached $1 million or more. Pennsylvania and New York are the two highest-awarding states.

Data on malpractice trends are largely provided by the insurance industry, which raises an issue of bias. A 2005 study collected more objective data from the National Practitioner Data Bank and from the Physician Insurers Association of America's Data Sharing Project. Based on their findings, the authors concluded that "... the total dollars in paid physician medical malpractice claims have approximately doubled in the past decade. Average defense costs per claim have increased substantially, also doubling ... the growth in dollars

---
[4] Eisenberg, D and Sieger, M. The Doctor Won't See You Now. *Time*, June 9, 2003, p. 55.
[5] Available at www.ama-assn.org. Accessed September 13, 2004.

paid on malpractice claims is mainly due to increases in the average size of claims. The total number of paid claims has been relatively stable, despite a sizeable increase in the number of physicians. The overall increase in total medical malpractice payments was only slightly above the rate of medical care inflation, but somewhat greater than the general rate of inflation."[6]

Malpractice litigation is not an entirely American phenomenon. Ireland and Australia, for example, are hotbeds of malpractice lawsuits as well. Even tiny Singapore, an independent and prosperous island-state of 4 million people, has witnessed a recent string of malpractice lawsuits that has led to an escalation in premiums. The number of claims in Singapore reported to MPS, its chief carrier, has risen 4-fold in the past decade, and recent cases have flirted with seven-figure damages, with the largest being $1.5 million dollars. However, the case was reversed on appeal.[7]

REASONS FOR THE MALPRACTICE CRISIS

Our system of personal injury litigation offers a huge incentive for an injured party to seek legal redress in the courts. Windfall profits can accrue to the plaintiff lawyer because of the contingency fee structure, wherein the attorney gets a percentage, usually a third, of any damages collected. The plaintiff incurs no attorney fees should the lawsuit fail. Moreover, unlike in the British system, the unsuccessful plaintiff does not even have to pay for the defendant's legal fees or the costs of expert witnesses. According to a legal scholar: *"Personal injury litigation is probably the most lucrative area of law in the United States — dwarfing even the returns from Wall Street firms ... personal injury lawyers share in the equity from huge verdicts and settlements, earning annual incomes that can approach or exceed $1,000,000."*[8]

The medical profession is adamant that there is no relationship between malpractice lawsuits and poor quality of care. That is, being sued for malpractice says nothing about a doctor's fitness to practice medicine. On the

---

[6] Budetti, PP and Waters, TM. Medical Malpractice Law in the United States. Publication #7328, available on the Kaiser Family Foundation's web site at www.kff.org.
[7] *Gunapathy Muniandy v. Dr. James Khoo et al.* Civil Appeal No. 600094 of 2001, Court of Appeal, Singapore.
[8] O'Connell, J. A Neo No-Fault Contract in Lieu of Tort: Pre-Accident Guarantees of Post-Accident Settlement Offers. *Cal L Rev* 1985; 73: 898, 903.

other hand, the plaintiff attorneys' bar has squarely asserted that malpractice suits exist for the simple reason that there is malpractice, i.e., sub-standard medical care.

So what are the reasons for mounting malpractice claims? It is unlikely that the general standards of medical care are slipping. The relentless increase in both frequency and especially severity of claims can be explained by one or more of the six reasons offered below. In the name of fairness, it should be acknowledged that some observers, trial lawyers among them, who believe there is nothing dysfunctional at all about our legal system of compensating injuries.[9] They prefer to characterize the current situation as a problem of insurance availability and affordability, laying the blame on the shoulders of the insurance industry for greed, poor investment performance, and vagaries of the economic cycle.

Reason #1. A Better Educated Public with Greater Awareness of the Medical and Legal Systems

No one knows for certain how often malpractice occurs. Everyone acknowledges that many meritorious claims are never pursued because of blind loyalty to the treating physician. This attitude is changing and the public is increasingly aware that doctors can act negligently and be held accountable for the harm they cause. Patients are shedding their respect of, and deference to, their treating doctors. They no longer accept the notion that "Doctor knows best," or attribute bad outcomes to bad luck or fate.

Better educated in medical matters, the public is more able and willing to question a doctor's diagnoses and treatments. Health-related information is widely available from many sources, with the Internet serving as an excellent repository, even if unfiltered and potentially misleading. The media play an important contributory role and are enthusiastic in their coverage of malpractice cases, drawing public attention to treatments allegedly gone wrong.

Affluence and education enable people to contemplate hiring lawyers to pursue legal redress for perceived wrongs.

---

[9]Association of Trial Lawyers of America, whose website is at atla.com.

Reason #2. Rising Expectations of Medical Results

The medical profession is rightly proud of its many achievements, and the treatment options available today are certainly more numerous and superior than in the past. We frequently trumpet our discoveries, sometimes prematurely, and there is an implied assurance that we can now treat most, if not all, diseases. Television depicts miraculous turnarounds. Patients sustaining cardiac arrests on programs such as *ER* are resuscitated and discharged alive and well, whereas in the real world only 10–15% of these patients survive to hospital discharge.

The profession and the media are inviting the public to expect — and demand — the very best treatment, the perfect cure. Anything short of that becomes a disappointment, which may breed dissatisfaction and anger. As one author put it: *"The promise of expertise must be kept. Humanity has split the atom and walked on the moon. Nothing less than a comparable performance in healthcare can be considered acceptable. Failure is no longer tolerated."*[10]

Reason #3. Medical Errors

A recent report from the Institute of Medicine estimates that almost 100,000 deaths occur annually as the result of medical errors. This report has generated much debate, and despite some criticisms, there is general agreement that medical errors do occur too often. Although medical errors are not exactly the same as malpractice, something ought to be done to reduce their frequency. The Institute of Medicine and the Joint Commission for the Accreditation of Healthcare Organizations (JCAHO) are leading this effort. Their mantra is to move "from a culture of blame to a culture of safety."[11]

The landmark study by Harvard investigators in the 1980s documents many examples of hospital errors, including negligent medical treatment, and this has led some to blame substandard practice as the root cause for rising malpractice premiums. This issue is contentious. Dr. Thomas Reardon, former

---

[10] Pollack, R. *Clinical Aspects of Malpractice.* Medical Economics Company, New Jersey, 1980, p. 3.
[11] Institute of Medicine: *To Err is Human: Building a Safer Health System.* National Academy Press, Washington D.C., 2000. The full text of the report is available online at www.nap.edu/readingroom. The Institute of Medicine's home page is at www.iom.edu.

AMA president, has remarked that *"Malpractice settlements are not a reflection of competency or quality... only one in five malpractice cases are due to negligence. The rest are due to our litigious society."*[12] His remarks are probably a fair representation of the medical profession's mindset on malpractice claims.

Reason #4. Commercialization of Medical Care with Erosion of the Doctor-Patient Trust Relationship

The doctor-patient relationship is founded on trust. Patients who have good relationships with their doctors are less likely to think of a lawsuit when there is an adverse outcome. However, today's practice climate requires the doctor to see more and more patients in less and less time. Additionally, because of cross-coverage, *locum tenens*, etc., the doctor-patient trust relationship may not be solidly formed. Increased specialization in medicine means the patient may not meet the doctor until there is a complication and the consultant, typically a total stranger to the patient and family, is brought in. Then there is the preoccupation with payment, and for hospitals, margin over mission. These shortcomings, coupled with failed communication and lack of informed consent, help to explain why patients are alarmed and angry at such an impersonal system that is more commercialized than caring.

Unfortunately, an adversarial attitude characterizes many doctor-patient encounters. Increasing distrust between the public and the medical profession is symptomatic of the changing attitudes and expectations of today's society, where interpersonal relationships seem to count for less.

Reason #5. Escalating Costs of Medical Care

Medical costs, including the cost of drugs, are spiraling out of control and patients are spending a greater portion of their budget on healthcare needs. The more one has to pay, the more intolerant one is of imperfection. This is true for a washing machine or hotel service; it is equally true for healthcare.

Reason #6. The Legal System of Litigating Injuries

Our system of litigating personal injury claims invites lawsuits because of two main defects: the contingency fee system, and astronomical jury verdicts.

---

[12] Kohl, M. New York Legislature Passes Physician Profiling Bill. *Dermatology Times* 2000; 21:16.

The first defect, a no-recovery-no-fee approach, frees the plaintiff from having to bear all legal costs, and may therefore encourage non-meritorious claims. Although many jurisdictions, including those of the British Commonwealth, equate the contingency fee system with unethical practice of law, this fee arrangement is enshrined in the U.S. legal system; it is also available in some European countries such as France, Slovakia, Estonia, Austria and Spain. The trial lawyers have borne the brunt of the harsh criticism for the second defect, a legal system "gone awry." Such disenchantment is widespread and politicized, and actually formed a part of the Republican Party's platform for the 2000 and 2004 presidential elections: *"Avarice among many plaintiffs' lawyers has clogged our civil courts, drastically changed the practice of medicine, and costs American companies and consumers more than $150 billion a year... This amounts to a tax on consumers to fatten the wallets of trial lawyers."*[13]

One author eloquently described it in this fashion: *"It is one of the most ubiquitous taxes we pay, now levied on virtually everything that we buy, sell and use. The tax accounts for 30 percent of the price of a stepladder and over 95 percent of the price of childhood vaccines. It is responsible for one-quarter of the price of a ride on a Long Island tour bus and one-third of the price of a small airplane. It will soon cost large municipalities as much as they spend on fire or sanitation services."*[14]

ADVERSE EFFECTS OF MALPRACTICE LITIGATION

Before identifying the multiple adverse effects from malpractice litigation, one must not lose sight of the fact that just compensation and deterrence of substandard conduct remain valid purposes for having the current legal system. Patients who have been injured through negligent care should be justly compensated. And there is evidence that the threat of being sued results in greater caution in the delivery of healthcare services. To the extent that this is kept in proper proportion, malpractice litigation improves the quality of care, and confers a benefit on the consuming public. On the other hand,

---

[13] At press-time, President Bush appears determined to introduce legislation aimed at rectifying aspects of our medico-legal system, specifically caps on non-economic losses and limitations on class-action lawsuits.
[14] Huber, PW. *Liability: The Legal Revolution and Its Consequences.* Basic Books, New York, 1988.

litigation run amuck leads to many bad outcomes, as follows:

Increased Healthcare Costs: Whenever healthcare providers pay their insurance premiums and/or malpractice damages, they ultimately pass on these costs to the consuming public. Some 70–80% of claims against physicians are terminated without indemnity payment. Still, in 2001, the cost of defending a non-meritorious suit averaged $23,000, and a verdict for the defendant at trial set the insurer back by $85,718.[15]

Moreover, malpractice fears result in the practice of defensive medicine, which is the process of conducting tests or treatment for legal rather than medical reasons. As one author has stated, defensive medicine encourages a physician to order $500 worth of tests that will give 98 percent diagnostic certainty, rather than $50 worth of tests that will yield 97 percent certainty.[16] In 1985, the AMA placed the annual cost of defensive medicine at $15.1 billion.[17] In that report, 40% of the responding physicians said they prescribed additional diagnostic tests and 27.2% said they provided additional treatment procedures as a response to the increased risk of a professional liability action. The Bush administration has estimated the Medicare price tag for perceived or real medical malpractice lawsuits at $25 billion annually.[18] And the most recent estimate of defensive medicine puts the figure at between $70 and $126 billion.[19]

Money is not the only issue. Defensive medicine carries with it a quantifiable risk of patient harm, since many medical tests pose material risks. Since virtually all tests carry at least some minor risk, proper medical decision-making requires that a test or treatment be ordered only when medically appropriate. Unfortunately, medico-legal considerations have crept into the definition of appropriateness.

Litigation Stress Syndrome: Even if the physician is not at risk for personal financial loss, the litigation experience regularly creates anxiety and a loss

---

[15] Anderson, RE. *Medical Malpractice: A Physician's Sourcebook.* Humana Press, New Jersey, 2005.
[16] Neubauer, D. Medical Malpractice Legislation: Laws Based on a False Premise (1985) *Trial* 64.
[17] American Medical Association Special Task Force on Professional Liability and Insurance. Professional Liability in the 80s, Reports 1, 2 & 3 (1984–5).
[18] Doctors' Activism Revives Malpractice Bill. *The Wall Street Journal*, January 13, 2003, p. A4.
[19] See AMA document entitled Medical Liability Reform-Now! Available at www.ama-assn.org/go/mlrnow.

of self-esteem. An adverse lawsuit outcome is also reportable to the State's licensing board and to the National Practitioner's Data Bank. These actions can threaten one's license to practice.

Some doctors are more likely than others to develop this litigation stress syndrome. In addition to the psychological trauma of a lawsuit and the threat to the doctor's livelihood, there is the unwelcome reminder each time the doctor faces the question: "Have you ever been sued for malpractice?" Such a question is repeated each time the doctor applies for new hospital privileges or re-credentialing, which is done every two years in most hospitals.

Physician fear of malpractice borders on paranoia. Even doctors who have never been sued are ever conscious of the sword of Damocles hanging over their heads. It has been said that many physicians live in an aura of fear — fear of suit. Medicine has been called the frightened profession, and physicians described as *"nude on the corner of the Main Street of life."* A malpractice lawsuit causes significant mental distress, which includes serious symptoms such as depression and suicidal ideations. Thirty-three percent of doctors thought about early retirement after being sued. One author noted: *"Any suit, even though frivolous, costs the defendant money, time, reputation, and peace of mind. It's the fact that suit was threatened or brought, rather than the jury's decision or the amount of the settlement, that concerns us most."* [20]

One study noted that more than one-third of all physicians who have been sued admitted to four or five symptoms suggestive of a possible major depressive disorder and 8% noted the onset of a physical illness. Some even succumb to suicide. Common symptoms are insomnia, anxiety, depression, anger, and diminished confidence and concentration.[21] Moreover, the doctor may experience decreased libido, guilt, preoccupation, helplessness and withdrawal from peers and family. It is the litigation process itself that is agonizing and stressful. Getting sued by patients is an event that has definite health repercussions.[22]

Physicians are vulnerable because they are expected to make no mistakes. Moreover, most doctors are demanding of themselves and hope to achieve perfection through diligence. They may feel that they have let the patient

---

[20] Pollack, R. *Clinical Aspects of Malpractice.* Medical Economics Company, New Jersey, 1980.
[21] American Medical Association Special Task Force on Professional Liability and Insurance. Professional Liability in the 80s, Reports 1, 2 & 3 (1984–5).
[22] Charles, SC and Kennedy, E. *Defendant: A Psychiatrist on Trial for Medical Malpractice.* The Free Press, 1985.

down or that they have been bad doctors. Physicians feel hurt and betrayed when they are sued. Many cannot confide in colleagues or family, and some may succumb to burnout or even resort to the use of drugs.

Loss of Medical Services: Another undisputed byproduct is physician frustration. Some retire early to escape the malpractice roulette, while others switch to less risky specialties or limit high-risk practice. A typical example is family practitioners who give up obstetric deliveries because of oppressive insurance premiums. The loss to society may be substantial; in certain locales, especially in rural areas, the net result may be the total loss of certain medical services. At the height of the malpractice crisis in the early 1980s, family practitioners on Molokai, a small island in the Hawaiian chain, stopped delivering babies because their premiums nearly doubled. A survey of Hawaii's physicians at the time found that 190 had taken, or were considering taking, early retirement.[23]

In parts of Mississippi, patients were forced to travel 65 mile for deliveries after a local hospital stopped offering its obstetric services.[24] In a 1999–2003 survey by the American College of Obstetricians and Gynecologists (ACOG), one in seven ACOG fellows have stopped delivering babies because of the fear of being sued.[25]

Obstetrics is not the only specialty affected. Some locales in West Virginia lost all neurosurgical services for two years because of unavailability of neurosurgeons, and a Level 1 trauma center in Nevada had to close for two weeks when 60 orthopedic surgeons halted services to protest soaring malpractice premiums.[26]

Erosion of Trust Relationship: Patients who trust their doctors derive maximum therapeutic benefit. An adversarial relationship between doctor and patient is inimical to the healing process. Mutual trust enables the doctor to elicit an accurate medical history and perform a proper physical examination. Medical advice for testing and treatment has to be given and accepted in an atmosphere of confidence and candor. Suspicion, hesitancy, or confrontation can irreversibly poison the relationship and thus impede the healing process.

---

[23] Hawaii Medical Association Newsletter, December 1985, p. 2.
[24] Unites States General Accounting Office. Medical Malpractice: Implications of Rising Premiums on Access to Health Care. Washington, D.C. 2003:29.
[25] Available at www.ama-assn.org. Accessed September 13, 2004.
[26] Unites States General Accounting Office. Medical Malpractice: Implications of Rising Premiums on Access to Health Care. Washington, D.C. 2003:29.

A significant part of the doctor's healing powers comes from the faith and trust of the patient. The power to reassure and convince is an integral part of the therapeutic armamentarium. The doctor must inform the patient of diagnostic and treatment plans — this is embodied in the doctrine of informed consent. The doctor should seek the patient's understanding and cooperation, and assuage the patient's fears. While the treatment goals sought by both patient and doctor are usually identical, the doctor must be allowed clinical judgment in decision-making. A solution to malpractice woes should jealously protect this implicit trust relationship. Sadly, doctors are now adopting a defensive posture towards their patients out of misplaced or exaggerated fear of a malpractice lawsuit.

Discouraging Graduates from Entering the Profession or Choosing High-Risk Specialties: There is a general belief that concerns over malpractice litigation discourage students from entering the profession. Even some doctors are encouraging their own children to look elsewhere. These are anecdotal observations, and the evidence indicates that applications for medical school are again on the way up (3.5% increase over previous year for the 2003 entering class).[27] However, the effect on the choice of high-risk specialties such as Obstetrics and Gynecology (ObGyn) is another matter. For the third year in a row, the number of medical students entering the ObGyn specialty has declined. In 2004, only 65% of the residency positions were filled with U.S. medical school graduates compared to 86% a decade earlier.[28]

NEW YORK AS AN EXAMPLE[29]

A Brief History: New York State has always been at the forefront of malpractice litigation and its various crises. Following a review of the Medical State Society of the State of New York's (MSSNY) publications from 1807 through the 1970s, Cirincione observed that the early twentieth century brought increasing concern of malpractice among physicians.[30] In 1959, R. Crawford

---

[27] Barzansky, B and Etzel, SL. Educational Programs in U.S. Medical Schools, 2003–2004. *JAMA* 2004; 292:1025–31.
[28] Available at www.ama-assn.org/ama/pub/article/9255-8713.html. Accessed September 13, 2004.
[29] Dr. Diane Pratt, Family Practice resident in Buffalo, assisted me in researching this section during a senior medical student elective.
[30] Cirincione, S. The History of Medical Malpractice in New York State — a Perspective from the Publications of the Medical Society of the State of New York. *N.Y. State J Med* 1986; 86:361–9.

Morris, a lawyer, published an article in the MSSNY journal entitled: *"Is there a doctor in the courtroom?"* in which he quoted a *Newsweek* article that *"since 1950 one out of every 35 doctors insured under the New York State Medical Society's group insurance plan has been sued in the courts for malpractice."*[31]

During the malpractice crisis of the 1970s, physicians joined forces and formed their own insurance companies; these were known as "bedpan mutuals." In June of 1975, *Newsweek* featured a lead article on medical malpractice, and reported on physician protests and demonstrations in New York and across the U.S. The article quoted Dr. Norman S. Blackman, president of the Kings County, N.Y. Medical Society, as saying: *"You do what's legally indicated, not what's medically indicated."*[32]

After a short lull with a somewhat more stable malpractice situation, the number of claims and in particular the size of the awards again soared. By the 1980s, the United States was facing its second medical malpractice crisis. Premiums skyrocketed and claim frequency rose again. The malpractice crisis of the 1980s witnessed the publication of the Harvard Medical Practice Study, which investigated the incidence of adverse medical events and negligence in New York's hospitalized patients. The study reviewed more than 30,000 randomly selected hospital records and 3,500 malpractice claims from New York in 1984. The authors found that 3.7% of hospitalizations resulted in adverse events, and 27.6% of these events were caused by negligence as determined by objective medical and legal reviews. The study revealed that only 1.53% of these injuries led to a malpractice claim. The conclusions were compelling: (1) malpractice litigation is an unjust mechanism for reimbursing patients victimized by medical negligence; and (2) there is no widespread proliferation of litigation over medical injuries.[33] Not surprisingly, the study has been criticized for its methodology and conclusions.[34]

The current situation: In 1999, the court's ruling in <u>Desiderio v. Ochs</u> served as a catalyst for the recent round of escalating malpractice premiums. The New York Presbyterian Hospital was charged with negligence because a child

---

[31] Cirincione at p. 368.
[32] Clark, M. Malpractice. *Newsweek*, June 9, 1975, p. 62.
[33] Brennan, TA *et al.* Incidence of Adverse Events and Negligence in Hospitalized Patients — Results of the Harvard Medical Practice Study I. *N Engl J Med* 1991; 324:370–6.
[34] Anderson, RE. An Epidemic of Medical Malpractice? A Commentary on the Harvard Medical Practice Study. The Doctors Company. Available at www.thedoctors.com. Accessed June 6, 2004.

suffered brain injury following surgery. Damages were to the tune of $80 million. The decision was affirmed by the Court of Appeals, with plaintiff agreeing to a reduction of damages to $50 million (before structured payments) in lieu of a new trial.[35] Cases like *Desiderios* led to several large insurance carriers leaving the market, including the well-known St. Paul Companies, which withdrew from New York in December 2001. St. Paul had provided coverage for roughly 40,000 physicians, 72,000 other healthcare professionals, and 750 hospitals and other healthcare facilities.[36] The companies that remained increased their premiums to meet rising costs and concerns over runaway jury awards. This made it difficult once again for physicians to obtain and afford coverage. Some carriers refused to cover new physicians or only accepted those with clean records.

The Medical Society of the State of New York (MSSNY) strongly agrees with the AMA that New York State is facing a malpractice crisis. Jury Verdict Research reported that the median award for medical liability cases rose from $500,000 to $1 million from 1995 to 2001. The largest single verdict in 2002, at $94,810,000, was awarded in Brooklyn.[37] New York was also mentioned in the "Top Ten Jury Verdicts of 2003" for its case, *Fellin v. Sahgal*. In a retrial, the jury awarded $112 million to a 23-year-old male who developed severe neurological injuries from a ruptured cerebral aneurysm.[38] He had arrived at the hospital complaining of back pain, severe headache and vomiting. The patient had no significant past medical history and initial vitals signs, including blood pressure, were within normal limits. The plaintiff alleged that the physician had failed to order a CT scan in a timely manner, thereby delaying surgery. In 2004, the verdict was set aside by the Kings County Supreme Court which ruled that there was "*no rational basis upon which the jury could have found that defendants' delay was a proximate cause of Fellin's injury.*"

The crisis has led to a reduction in services available to patients. Physicians have become reluctant to perform procedures such as obstetrics, and may refuse treatment for high-risk patients. This has further implications on patient access to appropriate healthcare. Women are known to drive over 50 miles for obstetrical care, and a New York City Council report observed

---

[35] *Desiderio v. Ochs*, 741 N.Y.S.2d 865 (2002); *affirmed* 791 N.E.2d 941 (N.Y. 2003).
[36] Hartwig R & Wilkinson C. Medical Malpractice Insurance. Insurance Information Institute 2003; 1:6.
[37] Civil Justice Reform. Medical Society of the State of New York, 2003.
[38] *Fellin v. Sahgal*, 745 N.Y.S.2d 565 (2002); *dismissed (JNOV)* 798 N.Y.S.2d 344 (2004).

that in certain areas, women must wait several months to receive a mammogram. In addition, the doctor-patient relationship has suffered as physicians subsequently scrutinize those patients who they think may be likely to sue.

The AMA quoted a second year medical student from Albert Einstein College of Medicine in New York: *"I'm not sure I'll be able to practice in New York or some of the other states I'm interested in living ... because the premiums are so high. I don't want to be nervous about practicing in an environment in which every possible move I make could lead to the end of my career as the result of an outrageous lawsuit."* [39]

Sandy Parker, President of the Rochester Business Alliance put it this way: *"Exorbitant malpractice insurance costs in New York State are a factor in preventing us from being able to attract the best and brightest physicians."* [40] Her concerns appear justified. In an American College of Obstetrics and Gynecology survey conducted in 2002, 16% of New York's obstetricians and gynecologists had stopped practicing obstetrics; 40% of the various counties had fewer than five practicing obstetricians; and as many as seven counties had no obstetricians. Furthermore, the Long Island Business News reported that 45% of obstetrics and gynecology graduates in 2002 had left the state.[41]

The shockingly high premiums in parts of New York is the likely explanation for the creation of underserved areas and the exodus of doctors, particularly obstetricians. Table 3 below shows that six-figure premiums are the rule for many surgical specialties, and according to the Insurance Advocate Albany Bureau, these can be expected to rise further.[42]

Table 3: Premiums (2003) in Metropolitan New York[43]

Neurosurgeon	Obstetrician	Orthopedic Surgeon
$157,000–$203,000	$95,000–$124,000	$71,000–$92,000

---

[39] Ben Galper, available at www.ama-assn.org.
[40] Underhill J: Medical Malpractice Tort Reform — How it Affects Insurers and Employers. *Business Strategies*, August 2003.
[41] *Long Island Business News*, March 28, 2003.
[42] Polovina, J. Med/Mal Premiums in New York Rising Average of 8.5%. *Insurance Advocate*, July 28, 2003, p. 7.
[43] Civil Justice Reform. Medical Society of the State of New York, 2003.

New York's Response: A recent survey reported that *"New Yorkers of all genders, regions, and political parties strongly support tort reform."*[44] The public agreed that there were too many lawsuits and wanted a cap on non-economic damages. Over 80% believed that the medical malpractice crisis was adversely affecting healthcare expenditures.

New York's response to the medical malpractice crises in the past has offered little relief. The American Academy of Actuaries reported that plans for reform were instituted in 1975, 1981, 1985, and 1986. These reforms did not include a cap on damages, and the tort reforms did not yield improvement.[45] In 1985, the state established the Excess Medical Liability Insurance Program as a quick fix; the intent of the program was to protect physicians whose coverage was inadequate. A supplementary amount of coverage was offered to physicians with hospital privileges and with a primary package of $1.3 million per lawsuit and a total of $3.9 million per year. According to MSSNY, 25,000 physicians have the excess coverage today. Although excess medical liability coverage has brought some relief to New York's unfortunate situation, the program still has its flaws. Interestingly, physicians are not required to carry liability insurance coverage in New York.[46]

New York already places a limitation on attorney fees, has a conditional periodic payment schedule, and requires notification of collateral payments to the court for their deduction from the total award (see Chapter 24 on Medical and Legal Reforms). New York does permit HMOs to contract with patients that all malpractice claims will be settled by arbitration. However, HMO members must have the option to refuse this contract. MSSNY Legislative Agenda for Tort Reform promotes several changes to the medical malpractice system.[47] These reforms include a $250,000 cap on non-economic damages; the opportunity for alternative dispute discussions of claims; a no-fault system for neurologically impaired infant cases; and a reduction in the periodic payment threshold. The agenda opposes legislation that removes limitations on attorney contingency fees, increases the amount of awarded damages in

---

[44] Survey: New Yorkers of All Genders, Regions, and Political Parties Strongly Support Tort Reform. The Business Council of New York State, Inc., April 2003.
[45] Medical Malpractice Tort Reform: Lessons from the States. American Academy of Actuaries, Fall 1996, pp. 3–4.
[46] New York Medical Malpractice Summary. McCullough, Campbell & Lane. Available at www.mcandl.com/newyork.html.
[47] Civil Justice Reform. Medical Society of the State of New York, 2003.

wrongful death cases, creates a date of discovery rule for the statute of limitations, and authorizes the addition of pre-judgment interest to an award that accrues prior to the announced verdict.

What crisis? There are some who deny that a liability crisis exists, or that the crisis is due to a dysfunctional legal system. Proponents of this view, most prominently trial lawyers, generally believe that medical errors and insurance greed are the two real reasons for escalating malpractice premiums. For example, the legislative director for New York Public Interest Group insists that *"physicians should focus on efforts to reduce medical errors, not reducing damages in malpractice lawsuits."*[48] And based on their studies showing that insurance premiums are influenced by the economic market cycle rather than the number of claims made, award size, or tort laws, the Americans for Insurance Reform rejected the notion that a crisis exists since its statistics indicated that the average claim award has remained constant between 1984 and 2001.[49]

Believing that public disclosure would deter medical negligence, New York Governor George Pataki signed the Physician Profiling Bill in the year 2000. The Bill allows for public access to the names, qualifications, and disciplinary actions of physicians who have been implicated in malpractice or punished by the medical board three times over ten years. This information is posted on the Department of Health's web site. New York is the 22nd state to offer this information for public knowledge. Needless to say, the AMA vigorously opposed the bill.

With such powerful polarizing views, dare we hope for anything better than a stalemate and a guarded prognosis?

---

[48] Kaiser Daily Health Policy Report Examines Medical Malpractice Developments in Eleven States. State Watch, February 20, 2004, p. 1.
[49] Medical Malpractice Insurance: Stable Losses/Unstable Rates in New York. Americans for Insurance Reform, January 2003, pp. 1–8.

SUMMARY POINTS

CONDITION CRITICAL, PROGNOSIS GUARDED

- Both the frequency and severity of medical malpractice claims continue to rise, and escalating insurance premiums once again plague the healthcare practitioner in the current malpractice crisis.

- Reasons for the crisis include increasing consumer awareness, unmet expectations, commercialization of medicine, escalating cost of medical services, and an ineffective or possibly faulty legal system.

- Malpractice litigation deters substandard care, but unbridled litigation increases societal costs and erodes the mutual trust between doctors and their patients.

2

What Is and Isn't Malpractice

A patient's request for treatment and the acceptance of a doctor's offered services with agreed upon fees for such services, have all the elements of a legally binding contract. When the treatment is not what was bargained for, the patient may bring an action for breach of this contractual agreement. However, over the years, an action in contract arising out of injuries received during medical care gradually evolved into one where there is a legal assessment of fault and harm. The law of torts became the governing law covering such injuries, although a claim based on breach of contract may still be made by the injured party especially when a particular result has been promised to the patient but not achieved.

The U.S. inherited the English common law, which are judge-made laws derived from cases decided by the courts of old England and following through to modern times. A prominent element of the common law is the doctrine of precedent or *stare decisis* (Latin for 'let the decision stand'). In other words, the legal principle laid down by an appeals court in a particular jurisdiction in an earlier case is binding on subsequent cases presenting with similar facts. If the legal issue is a novel one, the judge is then obliged to declare what the legal principle ought to be. Another way laws are created is through legislation, where the government. i.e., state legislature or Congress, enacts laws in areas as diverse as workers' compensation, organ donation, tort reform, and child abuse.

A tort is a civil wrong affecting private citizens that is not based on a breach of contract. The tort of negligence is a legal cause of action that covers conduct that does not meet the standard of a reasonably prudent person. Negligence has nothing to do with the good or bad intentions of the perpetrator, although there is a separate class of legal wrongs, termed intentional torts, where the wrongdoing is based on an intentional act. Examples of this latter category are assault and battery.

MALPRACTICE DEFINED

Malpractice is the tort of negligence that is committed by professionals such as physicians, dentists, engineers, and lawyers. Medical malpractice, also called medical negligence, is substandard conduct by a healthcare provider that causes harm to patients. There are four legal elements, listed in Table 4, that make up the tort of negligence:

Table 4: Legal Elements of Negligence

1. Duty
2. Breach of Duty
3. Proximate Causation
4. Damages

For a patient to recover damages for an alleged act of medical malpractice, he or she must prove all four elements. The level of proof is "more probable than not." Other synonyms are: Preponderance of evidence; more likely than not; balance of probabilities; greater than 50% likelihood. Note that in criminal cases, the prosecutor has to prove guilt of a crime beyond reasonable doubt, which is a heavier legal burden.

The tort of negligence begins with the finding that a legal duty is owed, such as within a doctor-patient relationship. The duty owed is that of appropriate care, and the plaintiff must prove with expert testimony that the defendant has breached this standard. What is this standard? It is identical in both American[50] and English[51] law. The American standard is best taken from Prosser's Textbook on Torts: "*The formula under which this usually is put to the jury is that the doctor must have and use the knowledge, skill and care ordinarily possessed and employed by members of the profession in good standing...*" The British standard was articulated in 1957 in the _Bolam_ case: "*the question to be asked when determining medical negligence is whether a*

---

[50] Prosser & Keeton on Torts, 5th edition, 1984, pp. 186–7.
[51] For a comprehensive review of the law in England, see *Principles of Medical Law* by I. Kennedy and A. Grubb, Oxford University Press, 1998.

doctor, in acting in the way he did, was acting in accordance with the practice of a competent, respected professional."[52]

Compensable injuries, either physical or mental, must have been caused by the act of negligence. This is termed proximate causation. If the healthcare provider is found liable, then he or she is required to compensate the injured patient and/or the family with a monetary sum called damages. Finally, even if the plaintiff has proven all four elements, the case may still be lost, or damages reduced, if the doctor can raise affirmative defenses such as contributory negligence by the plaintiff.

Malpractice is therefore an act, by commission or omission, by a healthcare professional that departs from an accepted healthcare standard, where the departure causes an injury to the patient. As articulated by the Supreme Court of Nebraska: *"In a malpractice action involving professional negligence, the burden of proof is upon the plaintiff to demonstrate the generally recognized medical standard of care, that there was a deviation from that standard by the defendant, and that the deviation was a proximate cause of the plaintiff's alleged injuries."*[53] This standard does not have to reach a high or an outstanding level, just the ordinary standard of the practitioner of that specialty, sometimes referred to as the 'minimum common skill.' As has been pointed out, it is not exactly correct to say 'average skills' since that implies that half of all doctors are below the mark (strictly speaking 'median' implies this distribution).[54]

It is also incorrect to say that medical negligence means an adverse outcome, a wrong judgment, or even a medical error or mistake. A medical error or misjudgment is not synonymous with medical negligence. It depends on the nature of the error or misjudgment. If it is one that a reasonably competent professional would not commit, then the standard of care is breached, and medical negligence has resulted. On the other hand, if reasonably skilled practitioners could commit such a misjudgment, then it would not amount to medical negligence. As one court put it: *"An honest error of judgment in making a diagnosis is insufficient to support liability unless that mistake constitutes negligence."*[55] Nor is an adverse outcome necessarily the result of

---

[52] *Bolam v. Friern Hospital Management Committee*, 1 WLR 582 (1957).
[53] *Hamilton v. Bares*, 678 N.W.2d 74 (Neb. 2004).
[54] Dobbs, DB. *The Law of Torts*, Chapter 14. West Information Publishing Group, 2000.
[55] *Dotson v. Hammerman*, 932 S.W.2d 880 (Mo. App. 1996).

negligence. Another court put it this way: *"The mere fact that the physician has failed to effect a cure or that the diagnosis and treatment have been detrimental to the patient's health does not raise a presumption of negligence."*[56] Some medical conditions end up with bad results that are wholly independent of the doctor's actions, and the doctor is neither an insurer nor a guarantor of the patient's health.

The Hawaii Supreme Court has ruled that use of the terms 'error in judgment' and 'best judgment' would confuse the jury, and re-emphasized the objective community standard against which medical negligence is to be measured.[57]

PROFESSIONAL MISCONDUCT

Professional misconduct is an ethical rather than a legal notion that speaks to unprofessional behavior; it extends beyond medical competency. The modern term is unethical conduct. This is not medical negligence, although ethical standards are also professional standards, and they may sometimes be used to define aspects of the standard of care. What constitutes professional misconduct is elusive; it is a British term, and the Council on Ethical and Judicial Affairs of the American Medical Association (AMA Council) does not use or define it. Professional misconduct, if proven, may be sanctioned in one of several ways, including probation and even loss of medical licensure. Thus, to be in breach of this ethical standard may be of greater concern to a doctor than to simply pay a sum of dollars in a malpractice suit. The rough equivalent of unethical behavior is 'infamous conduct in a professional respect,' a term taken from an old English case.[58] More recently, the term serious professional misconduct has replaced the earlier term 'infamous conduct' in a professional respect, with the unchanged definition of: *"serious misconduct judged according to the rules, written and unwritten, which govern the medical profession. If it is shown that a practitioner has done something in pursuit of his profession which would reasonably be regarded as disgraceful or dishonorable by his professional brethren of good repute and competence...."*[59]

---

[56] *Bryan v. Burt*, 486 S.E.2d 536 (Va. 1997).
[57] *Hirahara v. Tanaka*, 959 P.2d 830 (Haw. 1998).
[58] *Allinson v. General Council Medical Education and Registration*, 1 Q.B. 750 (1894).
[59] Brahams, D. The Meaning of Serious Professional Misconduct. *The Medico-Legal Journal* 1987; 55:3–5.

The AMA Council has described the most comprehensive set of norms and boundaries that govern the ethical conduct of doctors.[60] It lists nine guidelines in the preface, covering all aspects of medical practice. Other specialty organizations have also published ethical guidelines or codes of ethics, the best known being that of the American College of Physicians.[61]

CRIMINAL LIABILITY

A malpractice lawsuit is not a criminal matter. It is about compensating an injured party with a sum of money if legal fault can be proven. However, wrongdoing can amount to a crime where there is proven *mens rea*, which means the presence of criminal intent. Generally, this means acting purposefully, knowingly, or with reckless disregard or gross negligence.

Criminal prosecution of doctors for patient injuries or death is rarely sought by prosecutors. For example, California failed in its attempt to prosecute two doctors for withdrawing intravenous fluids in an irreversibly comatose patient that resulted in his death. The charge was homicide, but the Court decided that under the circumstances, there was no legal duty to treat.[62]

There has been the occasional case where a physician was criminally prosecuted and convicted. One such case is *People v. Einaugler*, where a New York doctor was convicted of willful neglect and reckless endangerment because he did not transfer a nursing home patient to a hospital in a timely fashion after he had learned of the patient's abdominal condition which turned out to be peritonitis. He was sentenced to serve 52 weekends in a New York prison. The New York Supreme Court held that the misdemeanor of willful violation of heath laws required showing of willful failure to provide treatment or care, not mere simple negligence. However the evidence was sufficient to establish the defendant's guilt of reckless endangerment in the second degree in connection with his failure to a timely transfer of the patient to a hospital.[63]

---

[60] American Medical Association Council on Ethical and Judicial Affairs, Code of Medical Ethics: Current Opinions with Annotations.
[61] *Ethics Manual, 5th edition*, American College of Physicians.
[62] *Barber v. Superior Court*, 147 Cal. App.3d 1006 (1985).
[63] *People v. Einaugler*, 208 A.D.2d 946 (N.Y. 1994).

Elsewhere, doctors have occasionally faced criminal charges for patient injuries. The classic English case is <u>Rex v. Bateman</u>. In that case, the doctor ruptured a woman's bladder, crushed her colon, and removed a portion of her uterus while attending to a difficult delivery that resulted in a stillbirth. Dr. Bateman was convicted and sentenced to 6 months' imprisonment, but was freed on appeal for technical reasons.[64] In another case, an anesthesiologist was convicted of manslaughter in the death of a patient undergoing surgery for a detached retina. During surgery, the patient's ventilation was interrupted because of accidental disconnection of the endotracheal tube for 4 minutes. This led to a cardiac arrest. An alarm had apparently sounded, but was not noticed. The injury would not have occurred had the doctor attended to the patient instead of being away from the operating room.[65]

A recent newsletter itemized several examples of criminal charges against physicians.[66]

- Murder charges against a Florida family physician for trafficking in controlled substances and improperly prescribing a painkiller that contributed to a young man's death.
- Negligent homicide and manslaughter charges against a Utah psychiatrist for over-medicating five elderly patients who subsequently died.
- Similar charges against a Colorado anesthesiologist who allegedly fell asleep during surgery and failed to respond to a patient's changing vital signs.
- Manslaughter charge against a Pennsylvania doctor for failing to properly examine a patient with intestinal obstruction. Four other doctors in the case were charged with assault and neglect.

---

[64] <u>Rex v. Bateman</u>, 19 Cr. App. Rep. 8 (1925).
[65] <u>R v. Adomako</u>, 2 All ER 79 (1994).
[66] Young, D. Criminalization of Medicine: Reducing the Risk. *Medical Liability Monitor*, July 2004; 29:8.

SUMMARY POINTS

WHAT IS AND ISN'T MALPRACTICE

- A tort is a civil wrong affecting private citizens that is not based on a breach of contract.
- The tort of medical malpractice, also known as medical negligence, is usually defined as an act or omission that breaches the standard of an ordinarily skilled practitioner, which results in injuries.
- Errors and misjudgments are not necessarily measures of legal negligence.
- Professional misconduct is not the same as malpractice and is more of an ethical breach than a legal issue.
- Occasionally, tortious conduct such as medical negligence can be a criminal matter that carries jail terms and/or monetary fines.

3

Legal Duty

The first element of the tort of negligence is the duty of reasonable care. To whom is this duty owed? Not everyone owes everyone else a duty. The usual test is whether the victim was foreseeable. In a famous New York case, a woman named Mrs. Palsgraf was standing on a platform under a scale in a New York train station, when a passenger dropped a plain-looking wrapped package as the conductor pulled him aboard the departing train. The package, which contained explosives, exploded and caused the scale to fall on Mrs. Palsgraf. Injured, Mrs. Palsgraf sued the railroad company but lost the lawsuit. Judge Benjamin Cardozo reasoned that there was nothing about the package that a reasonable 'person' (in this case, the railroad company) would anticipate should pose a danger to the victim. In other words, Mrs. Palsgraf was not a foreseeable victim, and, therefore, no duty was owed to her.[67]

In allegations of medical negligence, the doctor usually owes a duty to the patient who is making the claim. This duty arises out of the doctor-patient relationship, i.e., whenever a doctor undertakes to evaluate or treat a patient. In the absence of such a relationship, the law does not impose a duty of care. Generally, a doctor is not legally obligated to treat a total stranger who may be in need of medical assistance. However, his or her ethical, as opposed to legal, duty is a separate matter.

DOCTOR-PATIENT RELATIONSHIP

It may not always be obvious if a legal duty exists. Whether a doctor-patient relationship is formed is generally dependent on the particular facts of the case, but a good rule of thumb is that the more involvement a physician has with a patient, the more likely a doctor-patient relationship is formed.

---

[67] *Palsgraf v. Long Island R. R. Co.*, 162 N.E. 99 (Ct. App. N.Y. 1928).

Legal Duty

Suppose a doctor is attending to a patient and someone in the adjoining bed requires medical assistance. Does the doctor have a legal duty to respond? If a visitor passes out while visiting a sick relative, which doctor owes the duty to come to his/her aid? Depending on the facts, even some forms of curbside consults or medical advice given free-of-charge at social gatherings or over the phone or Internet may actually create a doctor-patient relationship. In one New York case, a physician speaking to a patient from the emergency department was deemed to have formed such a relationship.[68] Formal consults naturally do establish a physician-patient relationship, but even informal curbside consults may give rise to a legal duty. For example, an on-call neurologist's telephone consultation advice to an emergency room doctor raised the possibility that a relationship had been formed between the neurologist and the patient seeking treatment.[69] On the other hand, an informal consult from a consultant did not give rise to a relationship between the consultant and the patient even though the primary care doctor recorded the information in the medical record.[70]

What if a physician allowed his name to be entered as the attending doctor of record in order to facilitate a patient's hospital admission, and then informed the staff that he was not the treating physician? In one case under similar facts, the court refused to rule out the presence of a legal duty, and allowed the jury to decide.[71]

There has been a proliferation of litigation involving doctors who perform independent medical evaluations (IMEs) or employment physicals. Generally, there is no duty owed to the examinee in these situations, the doctor being responsible only for reporting the findings to the employer or insurer/attorney seeking the IME. The examinee will generally be unsuccessful if suing for a missed or wrong diagnosis. However, this rule is not absolute, especially where an unsuspected important medical condition is uncovered during the examination. Under these circumstances, it is prudent for the physician to advise prompt follow-up with his or her personal doctor. That the IME is not the equivalent of a regular checkup should be made clear to the 'patient' and

---

[68] *O'Neil v. Montefiore Hospital*, 11 A.D.2d 132 (N.Y.A.D. 1 Dept. 1960).
[69] *Lection v. Dyll*, 65 S.W.3d 696 (Tex. App.Dallas 2001).
[70] *Oliver v. Brock*, 342 So.2d 1 (Ala. 1976).
[71] *Giallanza v. Sands*, 316 So.2d 77 (Fla App. 4 Dist. 1975).

documented in the written report, which should be disclosed to no one other than the legitimate requesting party.[72]

GOOD SAMARITAN LAWS

Ordinarily there is no duty to aid a stranger in distress. However, to encourage such aid, so-called 'Good Samaritan' laws have been enacted that protect rescuers who act in good faith. These laws became popular in the 1960s and 1970s with the perception that doctors were reluctant to treat injured or ill strangers for fear of a malpractice lawsuit. California was the first state to enact such laws, and all other states have since followed its example.

Generally speaking, Good Samaritan laws immunize the aid-giver against allegations of negligent care. The actual scope of these protective statutes varies from state to state, but they do not protect against gross misconduct. California is an exception, as it may excuse even gross negligence so long as the act was done in good faith. In a litigated case, a California court declared that: *"The goodness of the Samaritan is a description of the quality of his or her intention, not the quality of the aid delivered."*[73] Acceptance or expectation of payment usually renders inapplicable Good Samaritan protection, and there may be differing standards for acts in the hospital setting, as opposed to those outside. Hawaii's law is typical, and portions of its Good Samaritan statute[74] are reproduced below:

> (a) Any person who in good faith renders emergency care, without remuneration or expectation of remuneration ... shall not be liable for any civil damages resulting from the person's acts or omissions, except for such damages as may result from the person's gross negligence or wanton acts or omissions.
> (b) No act or omission of any rescue team or physician working in direct communication with a rescue team operating in conjunction with a hospital or an authorized emergency vehicle ... shall impose any liability ... if good faith is exercised.

---

[72] Baum, K. Independent Medical Examinations: An Expanding Source of Physician Liability. *Ann Int Med* 2005; 142:974–8.
[73] *Perkins v. Howard*, 232 Cal.App.3d 708 (1991).
[74] Hawaii Revised Statutes §663-1.5.

> (c) Any physician ... who in good faith renders emergency medical care in a hospital ... shall not be liable ... if the physician exercises that standard of care expected of similar physicians under similar circumstances.

Good faith is defined in this Hawaii law as "*a reasonable opinion that the immediacy of the situation is such that the rendering of care should not be postponed.*" Texas has a similar statute which provides, in part:

> *(a) A person who in good faith administers emergency care, including using an automated external defibrillator, at the scene of an emergency but not in a hospital or other healthcare facility or means of medical transport is not liable in civil damages for an act performed during the emergency unless the act is willfully or wantonly negligent.*[75]

Note that Hawaii and many other states impose the simple negligence rather than the gross negligence or reckless disregard standard for Good Samaritan acts within a hospital. In other words, there is no additional statutory protection for being a Good Samaritan doctor in a hospital setting. Outside a healthcare setting, good faith and lack of gross negligence will confer immunity. California goes even further. Its statute virtually eliminates all liability for Good Samaritan acts, even for gross or wanton negligence, so long as the practitioner acts in good faith.

Commentators have observed that in fact, very few lawsuits have involved Good Samaritan doctors and that such laws are both unnecessary and ineffective. Those who are averse to helping will remain on the sidelines even with the protection of the law. The doctor working in an emergency situation is unlikely to provide anything more than first-aid or CPR, and without the proper diagnostic tools, equipment, drugs, etc., will likely be judged by a lower standard of care. Finally, it must be stressed that the ethical responsibilities of a doctor, which regularly exceed legal obligations, may morally mandate the provision of emergency medical care irrespective of whether there is a prior established doctor-patient relationship. The American Medical Association's Code of Medical Ethics states that: "*Physicians are free to choose

---

[75]Texas Good Samaritan Act, Civ. Prac. & Rem. §74.001.

whom they will serve. The physician should, however, respond to the best of his or her ability in cases of emergency where first aid treatment is essential."[76]

EMTALA

Healthcare facilities such as hospitals or clinics can generally choose their patients so long as they do not violate anti-discriminatory civil rights laws. One major exception, however, occurs in the evaluation and treatment of patients in an emergency. In 1986, Congress enacted a law called the Emergency Medical Treatment and Active Labor Act (EMTALA) which forces participating facilities that accept Medicare reimbursement to provide appropriate medical screening and stabilization of patients who present to the emergency department. The intent of the law is to stop 'dumping,' i.e., the shunting of non-paying poor and uninsured to other hospitals.[77] At a minimum, such patients must be stabilized before transfer, if indeed full and proper treatment facilities are not available.

EMTALA is a federal statute covering emergency services. It is not a general malpractice statute, and does not preempt state tort law claims.

DUTY TO THIRD PARTIES

Sometimes, a doctor is liable for someone other than his or her immediate patient. In such a circumstance, another person, often referred to as a 'third party,' may sue the doctor absent a doctor-patient relationship. An example is where an obstetrician fails to treat a pregnant woman known to have been exposed to German measles, who then delivers a child with birth defects. A Rhode Island court has ruled that a cause of action could be instituted by the child, the so-called 'third party.'[78] In another case, a missed diagnosis of meningitis in a mother led to the son contracting and dying from the disease. The lower court found no physician-patient relationship between the doctor and the son, but the appellate court reversed, holding that the physician-mother relationship resulted in a special situation for imposing a duty of care for her son.[79] Similarly, the Supreme Court of Tennessee held that a physician has a duty to warn persons in the patient's immediate family of the risk of

---

[76] AMA Code of Medical Ethics §8.11, 2000–2001 edition.
[77] 42 U.S.C.A §1395dd.
[78] *Sylvia v. Gobeille*, 220 A.2d 222 (R.I. 1966).
[79] *Shepard v. Redford Community Hospital*, 390 N.W.2d 239 (Mich. App. 1986).

a disease such as Rocky Mountain Spotted Fever although it itself was not contagious.[80]

A doctor may even be found to have a duty to a total stranger. The best known case is _Tarasoff_,[81] where a California court imposed a duty on a psychologist to warn an intended named victim even though that meant breaching confidentiality of a professional relationship (see Chapter 17: Medical Records and Confidentiality).

An emerging area of malpractice litigation affects patients who drive. Consider the following example: Suppose a patient drives his car and hits a pedestrian. The patient is a diabetic and was recently placed on insulin. The accident was caused by the loss of control of his vehicle because of hypoglycemia. The pedestrian could bring a lawsuit against the driver, but could also bring a suit against the doctor for failing to adequately warn of the risk of hypoglycemia, its prevention, and its treatment. The doctor is now faced with a potential liability to the injured pedestrian, a total stranger. Such an issue, with slightly different facts, was recently decided by the Hawaii Supreme Court which held that: *"A physician owes a duty to non-patient third parties injured in an automobile accident caused by an adverse reaction to the medication... where the physician has negligently failed to warn the patient that the medication may impair driving ability and where the circumstances are such that the reasonable patient could not have been expected to be aware of the risk without the physician's warning."*[82] The medication in this case was the anti-hypertensive drug Prazosin or Minipress (See Chapter 11: Patients Who Drive: A New Worry for Doctors).

ABANDONMENT

The flip side of duty is abandonment, which is recognized as both a legal wrong and an unethical act. However, a doctor is permitted to unilaterally terminate the professional relationship with a patient, but it must be done in a manner that does not result in harm. Where there is a prior established doctor-patient relationship, the doctor cannot unilaterally abandon the patient, e.g., withdrawing from treating a pregnant woman, or refusing to make a referral or provide timely and necessary information to a new treating doctor.

---

[80] _Bradshaw v. Daniel_, 854 S.W.2d 865 (Tenn. 1993).
[81] _Tarasoff v. Regents of University of California_, 551 P.2d 334 (Cal. 1976).
[82] _McKenzie v. Hawaii Permanente Medical Group, Inc._, 47 P.3d 1209 (Haw. 2002).

The tort of abandonment may be intentional or negligent. In the former instance, the deliberateness of the defendant's action is a key element, and the aggrieved party need not produce expert testimony to prevail. Intentional abandonment may be expressed or implied. On the other hand, negligent abandonment, e.g., premature discharge of a patient from the hospital or emergency department, is simply a form of medical negligence which will require expert testimony at trial to set the standard of care.

When a physician wishes to sever the professional relationship because of disagreements over treatment plans, or loss of rapport or trust, he or she has to give sufficient notice to the patient, and arrange to transfer all medical records to the new treating doctor. In addition, the original physician must be available to provide all emergency care until the patient has seen a new doctor. In other words, it is the affirmative obligation of the physician to ensure that no interim harm comes to the patient whom he or she is discharging. Merely making a referral may not be enough. In a dental malpractice case where the patient was simply referred to his family physician after oral surgery was complicated by an infection, an Illinois court held the dentist liable for the post-op complications.[83]

A not uncommon scenario is where a group practice collectively refuses to treat a patient or group of patients. In *Leach v. Drummond Medical Group, Inc.*, the plaintiffs complained to the State licensing board about some members in a group practice, and the group then attempted to terminate its professional relationship with those patients. The California appellate court held that whereas one member may decline to treat, the entire group may not exercise the same option, as there were no similar services within a hundred miles.[84]

REFUSAL TO TREAT

On a few rare occasions, the facts were so outrageous that the courts ruled that the healthcare providers owed no further duty to provide treatment.[85] The best known case is that of *Payton v. Weaver*, where a physician and hospital refused to provide dialysis to a patient because of persistent uncooperative

---

[83] *Longman v. Jasiek*, 414 N.E. 2d 520 (Ill. App. 1980).
[84] *Leach v. Drummond Medical Group, Inc.*, 144 Cal. App.3d 362 (1983).
[85] Orentlicher, D. Denying Treatment to the Noncompliant Patient. *JAMA* 1991; 265:1579–82.

and antisocial behavior. In lamenting that *"occasionally a case will challenge the ability of the law, and society, to cope effectively and sensitively with fundamental problems of human existence,"* the Court affirmed the trial court's judgment that the healthcare providers had no legal obligation to continue providing regular dialysis treatment.

The facts of the case are as vivid as they are tragic. Ms. Brenda Payton was a 35-year-old woman with end-stage-renal disease on dialysis. An active user of alcohol, heroin and barbiturates, she was non-compliant with dialysis and medical therapy. She continued to buy barbiturates from pushers on the street at least twice a week, failed to restrict her diet, sometimes gaining 15 kg. between dialysis, and frequently missed dialysis, resulting in 30 emergencies requiring hospitalizations in the eleven months before trial. She would come to dialysis drugged or drunk, use profane and vulgar language, cuss staff with obscenities, and expose her genitals in a lewd fashion. At times, she would scream for dialysis to be terminated, and pull the dialysis needle from the shunt causing blood to spew.

The Court found her behavior to be *"knowing and intentional"* and described Dr. Weaver as *"one of the most sensitive and honest physicians that I have been exposed to either in a courtroom or out of a courtroom."* It held that *"there exists no basis in law or in equity to saddle him with a continuing sole obligation for Brenda's welfare."*[86]

In another refusal-to-treat case, an abusive and disruptive patient missed scheduled dialysis sessions, and threatened to kill his nephrologists, Dr. John Bower, and shoot the hospital administrator at the University of Mississippi Medical Center. The Court ruled that to compel Dr. Bower to proffer dialysis was tantamount to involuntary servitude, and this would violate the 13th amendment which prohibits slavery. However, it ordered the medical center to continue to provide dialysis because there was no other facility in the region.[87]

---

[86] *Payton v. Weaver*, 131 Cal.App.3d 38 (1982).
[87] *Brown v. Bower*, No. J86-0759 (B), SD. Miss. Dec. 21, 1987.

SUMMARY POINTS

LEGAL DUTY

- Doctors owe a legal duty of reasonable care to their patients but not to strangers. However, doctors have an ethical duty to treat strangers in grave distress.
- Good Samaritan statutes attempt to immunize voluntary helpers from a negligence claim for harm done while rendering help in good faith and without compensation.
- Occasionally, a duty of care may also be owed to a third party.
- Doctors cannot abandon their patients. To terminate a relationship, doctors must give notice and ensure interim care until the patient is transferred to a new doctor.

4

Standard of Care

The previous chapter addressed the question, "To whom does the healthcare provider owe a duty?" This chapter asks "What kind of duty is owed?"

In negligence law, the duty that is owed is the duty of due care, or what a reasonably prudent person would do under the circumstances. Stated more formally, negligence is *"conduct which falls below the standard established by law for the protection of others against unreasonable risk of harm."*[88] The standard is, therefore, an objective one — as judged by the reasonably prudent person, which means the jury. If the conduct of the tortfeasor, i.e., wrongdoer, falls below this standard, then a breach of duty is said to have occurred.

This standard is somewhat modified for medical professionals who are alleged to have caused injuries to patients. It has long been recognized that the average layperson was incapable of judging what the acceptable level of medical care ought to be. The law, therefore, has taken the position that the standard is that level of care expected of the reasonably competent doctor, rather than the reasonably prudent person. Alabama, for example, has held that physicians must *"exercise such reasonable care, diligence, and skill as reasonably competent physicians"* would exercise in the same or similar circumstances.[89] An Illinois court used similar words: *"[a] physician must possess and apply the knowledge, skill, and care of a reasonably well-qualified physician in the relevant medical community."*[90]

Occasionally, injuries sustained in a healthcare setting may be judged by the reasonably prudent person standard. While the professional standard pertains to injuries arising out of healthcare, the reasonable person standard

---

[88] Restatement (second) of Torts §282 (1965).
[89] *Keebler v. Winfield Carraway Hospital*, 531 So.2d 841 (Ala. 1988).
[90] *Purtill v. Hess*, 489 N.E.2d 867 (Ill. 1986).

governs non-healthcare activities such as falls on slippery hospital floors. Unfortunately on occasion the distinction is unclear. As one author put it, *"Sometimes it is difficult to differentiate bad housekeeping and bad medical care, as where rats in a hospital repeatedly bit a comatose patient."*[91]

LOCALITY RULE

In the past, courts would use the standard of the particular locale where the tortious act took place, invoking the so-called 'locality rule.' This was based on the belief that different standards of care were applicable in different areas of the country, e.g., urban or rural. However, this rule has been largely abandoned in favor of a uniform standard, because medical training and board certifications all adhere to a national standard. Telemedicine has further propagated this uniformity.

With the erosion of the locality rule, courts now readily allow out-of-state experts to testify on behalf of the opposing parties. This has been especially helpful for plaintiffs who are far less likely to be able to secure willing experts from the local community.

Geographical considerations are not entirely irrelevant. Where the local medical facilities lack state-of-the-art equipment or specialists, courts will look at the existing circumstances. However, there is always the duty to refer and transfer to an available specialist, and the failure to do so may form the basis of liability.

SETTING THE STANDARD OF CARE

An allegation of malpractice is not about the physician's bad judgment, bad faith, or intentional malfeasance. It is about breaching an objective standard of medical practice. As a rule, expert testimony is required to establish the custom of the profession. Both the complaining patient and the defendant doctor are required to produce experts to legally establish what constitutes standard as opposed to substandard care. Experts, by virtue of their skills, knowledge, experience or education — supported by authoritative texts and

---

[91] Dobbs, DB. *The Law of Torts*, Chapter 14. West Information Publishing Group, 2000, referring to *Lejeunee v. Rayne Branch Hospital*, 556 So.2d 559 (La. 1990).

treatises as necessary — then articulate the standard as it applies to the particular case. In reaching their verdict, the jurors listen to all the evidence and decide which expert, and therefore which of the parties, is the more credible (see Chapter 7: Expert Testimony).

A plaintiff who attempts to sue a practitioner without the assistance of an expert witness is likely to have the case thrown out at an early stage. The law generally prohibits a lay person from setting the standard of care in professional negligence disputes; the plaintiff cannot simply refer to a book or article to support his or her case. In many ways, therefore, this is the biggest stumbling block for the plaintiff, as it is not always an easy or inexpensive matter to secure the services of an expert witness who is willing to testify against a doctor. Experienced plaintiff law firms, however, count on reliable medical contacts to review the case.

In legal proceedings addressing the standard of care, the doctor is judged according to his or her specialty. A general practitioner (GP) will not be held to the same standard of care as a specialist. The surgeon will be judged according to the community standard of the ordinarily skilled surgeon, and the GP to that of his fellow GPs. But there is a separate duty to refer to a specialist if the case is outside the doctor's field of expertise. If the standard of care is to refer to a specialist, the GP who undertakes to treat the patient within that specialty will be held to that higher standard. In *Simpson v. Davis*,[92] a general dentist performed root canal work and was therefore held to the standard of an endodontist.

Inexperience is not a defense. This seems particularly harsh to the trainee who cannot be expected to perform at the level of a fully trained or experienced practitioner. Yet, the trend is to hold medical trainees to the same standard as a qualified doctor in that specialty (see Chapter 16: Medical Trainees).

RESPECTABLE MINORITY RULE

The standard of care does not have to be a unanimous community standard. In medicine, there is frequently a minority view of how things ought to be

---

[92] *Simpson v. Davis*, 549 P.2d 950 (Kan. 1976).

done. So long as this minority view is held by a respectable group of doctors, the law will accept it as a legitimate alternative.

What can this mean in practice? If an expert presents a 'respectable minority' view, then the court or jury may find in favor of the defendant even if there are other acceptable ways of diagnosing or treating the patient. However, this does not mean that any 'on-the-fringe' publication on an issue will rise to the level of the respectable minority. Even if the information is allowed into evidence, the jurors may choose not to accept it.

A minority view may simply be considered reflective of a different approach to the same problem, but the care rendered must still comply with the standard of care of the minority view. In one Texas case, the court was not concerned with whether the practice was that of a 'respectable minority' or a 'considerable number' of physicians, but whether it met the standard. The case involved an augmentation mammoplasty procedure that resulted in silicone leakage. A number of qualified physicians had used that procedure, and this satisfied the court that the standard had been met.[93]

PHYSICIANS' DESK REFERENCE (PDR) AND MANUFACTURERS' PACKAGE INSERTS

The PDR lists all approved prescription drugs and provides detailed information covering, among other things, indications, contraindications, side-effects and warnings. Manufacturers of medical devices or equipment likewise provide detailed information governing the use of their products. These instructions are termed 'package inserts.' Some courts have chosen to accept package inserts as setting the standard for drug or device use, and will hold a physician liable if it can be shown that the product used by the patient was contrary to issued instructions. This approach was adopted by a Minnesota court in a wrongful death claim against a physician for using repeated doses of the drug chloromycetin to treat an ear infection. The patient died from aplastic anemia, a known complication of the drug. The court held that "*drug manufacturer's instructions concerning use of its product governed profession in prescribing treatment . . . sufficient to make out a prima facie case of liability.*" This is now widely known as the Mulder Rule.[94]

---

[93] *Henderson v. Heyer-Schulte Corp. of Santa Barbara*, 600 S.W.2d 844 (Tex Civ. App. 1980).
[94] *Mulder v. Parke Davis & Company*, 181 N.W.2d 882 (Minn. 1970).

Other jurisdictions do not go as far, and merely accept the PDR as evidence to be considered. Hawaii is a jurisdiction that rejects the Mulder Rule. In its seminal case on the issue, a 21-year-old woman underwent breast augmentation surgery, and subsequently noticed hardening of the right breast. Dr. Peebles, the defendant, diagnosed capsular contracture and performed closed capsulotomy, which unfortunately led to implant rupture. The plaintiff argued that the 'common knowledge' exception transformed her malpractice lawsuit into an ordinary negligence case, thus obviating the necessity of expert testimony to establish the applicable standard of care. She asserted that the 'common knowledge' exception applied because of specific information contained in the manufacturer's package insert.

However, the Hawaii court held that *"a manufacturer's package insert, in and of itself, may not establish the relevant standard of care in a medical negligence action. The inserts may be considered by the fact finder along with expert testimony, but may not alone define the standard of care."* It drew attention to the fact that the American Medical Association, while recognizing package inserts as one useful source of information, has maintained that inserts are an inadequate standard for medical practice, as they serve inconsistent purposes such as advertising for the manufacturer, regulation by the government, and information for the doctor.[95]

This same position was taken by a California Court in the case of a patient who sustained spinal cord injury following the injection of Sodium Urokon during translumbar aortography. The manufacturer's package insert recommended 10–15 ml, whereas the defendant used the customary dose of 50 ml. The plaintiff asserted that this spoke for itself as proof of negligence, but the Court held otherwise: *"Thus, while admissible, it cannot establish as a matter of law the standard of care required of a physician in the use of the drug. It may be considered by the jury along with the other evidence in the case to determine whether the particular physician met the standard of care required of him."*[96]

Doctors sometimes treat a condition with drugs that have not been approved by the FDA for use in that specific condition. This is called 'off-label'

---

[95] *Craft v. Peebles*, 893 P.2d 138 (Haw. 1995).
[96] *Salgo v. Leland Stanford Jr. Univ. Board of Trustees*, 317 P.2d 170 (Cal. 1957).

use. Whether such 'off-label' use amounts to negligence will depend on the community standard of practice as articulated by medical experts.

PRACTICE GUIDELINES

In recent years, various medical specialty organizations and governmental and commercial enterprises have issued practice guidelines that purport to define the best evidence-based medicine. The courts have tended to use these guidelines as reflective of current medical standards because they are usually arrived at by consensus of an objective authoritative body of clinicians such as the American College of Surgeons.[97] Some states such as Maine have passed legislation that allows doctors to elect to be covered by practice guidelines, with such compliance constituting evidence against an allegation of negligence.[98] Kentucky's statute presumes that the doctor has met the appropriate standard of care when the treatment has been in compliance with these guidelines.[99] On the other hand, other states such as Maryland, have ruled that practice guidelines are inadmissible as evidence in courts of law.[100]

EXCEPTIONS TO THE NEED FOR EXPERT TESTIMONY

An occasional case of medical negligence may not need the testimony of an expert to establish the standard of care. The defendant's own testimony can sometimes be sufficient. Criminal defendants have a fifth Amendment right against self-incrimination, but civil defendants do not. Theoretically, the plaintiff can call the defendant to state the requisite standard of care, but it is unlikely that the testimony will be what the plaintiff seeks given that the defendant is an 'adverse or hostile witness.' The two most common exceptions to the requirement for expert testimony are *res ipsa loquitur* and violation of statutes.

Res Ipsa Loquitur: The doctrine of common knowledge, more technically called *res ipsa loquitur* or 'the thing speaks for itself,' holds that where "*the plaintiff's evidence of injury creates a probability so strong that a lay juror*

---

[97] *Pollard v. Goldsmith*, 572 P.2d 1201 (Ariz. App. 1977).
[98] Me. Rev. Stats. Ann. §2975.
[99] Ky. Rev. Stat. §342.035.
[100] Ann. Code Md., Health-Gen. §19-1606.

can form a reasonable belief," a plaintiff may be entitled to a waiver of the requirement of expert testimony.[101] This doctrine is rarely invoked, usually in obvious examples of medical injuries such as amputation of the wrong limb, lung puncture following routine shoulder injection, or removal of the wrong vertebral disc. In one not so obvious example, the court allowed the case to go to the jury without the benefit of expert testimony on the basis of common knowledge. The case involved the severance of a patient's ureter during a complicated hysterectomy.[102] On the other hand, an Illinois court disallowed a plaintiff from claiming that it was common knowledge that someone should be referred to a cardiologist for a heart condition.[103]

Res ipsa loquitur had its genesis in the classic 1863 English case where a barrel of flour fell upon the plaintiff from a window above a shop. Despite no other evidence, the Court ruled for the plaintiff, opining that the circumstances constituted *prima facie* evidence of negligence (A *prima facie* case means the plaintiff has met the burden of going forward with evidence on the legal issue):

> *"I think it apparent that the barrel was in the custody of the defendant who occupied the premises, and who is responsible for the acts of his servants who had control of it; and in my opinion the fact of its falling is prima facie evidence of negligence..."*[104]

The *res ipsa loquitur* doctrine is most useful when the plaintiff has insufficient evidence of what caused the negligent act, but circumstances clearly indicate that the defendant was negligent. It is applicable only when three conditions are met:

(1) The event, under the circumstances of the case, ordinarily does not occur in the absence of someone's negligence.
(2) The event must be caused by a means within the exclusive control of the defendant.
(3) The plaintiff did not contribute to the event.

---

[101] *Gordon v. Glass*, 785 A.2d 1220 (Conn. App. 2001).
[102] *Cangemi v. Cone*, 774 A.2d 1262 (Pa. Super. Ct. 2001).
[103] *Evanston Hospital v. Crane*, 627 N.E.2d 29 (Ill. App. 1993).
[104] *Byrne v. Boadle*, 2 H. & C. 722, 159 Eng. Rep. 299 (Court of the Exchequer 1863).

In most jurisdictions, *res ipsa* permits the jury to infer that a negligent act had taken place, but the defense may still be able to rebut the evidence. Courts are usually hostile to the use of the *res ipsa* doctrine in medical malpractice, unless the circumstances clearly warrant the application of the doctrine such as in the case of foreign bodies that are left within body cavities following surgery. On the other hand, dysuria in association with a deformed penis was not sufficient evidence by itself to indicate negligent circumcision.[105]

In the well known California case of *Ybarra v. Spangard*, the court permitted the use of the *res ipsa* doctrine against multiple defendants in the operating room after the plaintiff developed shoulder injuries following an appendectomy.[106] Since the plaintiff was unconscious, the Court felt it was appropriate to place the burden on defendants to explain how the shoulder injury occurred. The *res ipsa* doctrine was also allowed in a case where the plaintiff sustained injuries to the peroneal and tibial nerves after knee surgery.[107]

Courtroom eloquence concerning *res ipsa* was at its best in *Cassidy v. Ministry of Health*, an English case from the 1950s. In *Cassidy*, a patient suffered significant deformity of his hand following surgery for Dupuytren's contracture. His attorney asserted: *"At the outset, only two of the plaintiff's fingers were affected; all four are now useless. There must have been negligence — res ipsa."* The Court of Appeal agreed, Lord Denning taking the position that it raised a *prima facie* case against the hospital. However, Lord Denning also indicated that the doctrine could only be invoked against a doctor in extreme cases.[108]

The use of *res ipsa* is governed by statutes in some states. Georgia and North Dakota, for example, disallow the use of *res ipsa* in medical negligence cases.

Violation of Statutes: If a healthcare provider violates a statute or regulation, and a patient incurs harm from this violation, the patient may have a good case to assert a breach of duty. The conduct has to affect the class of victims of which the plaintiff is a member, and be the type of injury that the statute was intended to protect. An example is where negligence is alleged against

---

[105] *Walker v. Skiwski*, 529 So.2d 184 (Miss. 1988).
[106] *Ybarra v. Spangard*, 154 P.2d 687 (Cal. 1944).
[107] *Hale v. Venuto*, 137 Cal.App.3d 910 (Cal.App. 1982).
[108] *Cassidy v. Ministry of Health*, 2 KB 343 (1951).

a practitioner without a valid medical license. Court opinions have differed, depending on whether the license had innocently lapsed,[109] or whether the practitioner was unqualified.[110]

In some jurisdictions, violation of a statute can amount to negligence *per se* (also termed statutory negligence), which means that there is no need for expert testimony. However, the plaintiff will still need an expert witness to prove causation. An example is the case of Landeros v. Flood. In Landeros, an emergency room physician failed to diagnose battered child syndrome, and discharged the child who subsequently suffered additional injuries at home. An existing statute mandated reporting of child abuse, and the doctor's breach of his statutory duty created a basis of liability for the missed diagnosis and injuries.[111]

If a statutory violation amounts to a rebuttable presumption of negligence, this may allow the plaintiff to get to the jury which can then accept or reject the presumption of fault.[112]

JUDGE-MADE STANDARDS

Medical standards are issues of fact that are ultimately determined by the jury after listening to the experts. It is rare therefore for a judge in a jury trial to decide on what constitutes the proper community standard. But in 1974, the Supreme Court of Washington did just that.[113] It held, as a matter of law, that tonometry or the measurement of intraocular pressure to diagnose glaucoma should be performed on all patients regardless of age. The standard of care at that time was to obtain such measurements only in those past the age of 40 because glaucoma is rare in younger patients. The case involved a 32-year-old woman who became blind because of the failure over a five-year period of various treating ophthalmologists to measure her intraocular pressures. The Court decided that it would institute its own standard in the name of public safety, since tonometric measurements are easy to perform

---

[109] McCarthy v. Boston City Hospital, 266 N.E. 2d 292 (Mass. 1971).
[110] Stahlin v. Hilton Hotels Corp., 484 F.2d 580 (7th Cir. 1973).
[111] Landeros v. Flood, 551 P.2d 389 (Cal. 1976).
[112] Martin v. Herzog, 228 N.Y. 164 (1920).
[113] Helling v. Carey, 519 P.2d 981 (Wash. 1974).

and may be sight-saving:

> "Courts must in the end say what is required; there are precautions so imperative that even their universal disregard will not excuse their omission ... Under the facts of this case reasonable prudence required the timely giving of the pressure test to this plaintiff. The precaution of giving this test to detect the incidence of glaucoma to patients under 40 years of age is so imperative that irrespective of its disregard by the standards of the ophthalmology profession, it is the duty of the courts to say what is required to protect patients under 40 from the damaging results of glaucoma."

Some seven years earlier, the same court had held that to permit a surgical operation in an anesthetized patient without a supervising doctor in the operating room amounted to "*negligence as a matter of law.*"[114]

There has not been a proliferation of cases where judge-made standards supplanted the traditional medical standard of care established by expert testimony. Allowing judicial weighing of risks versus utility was however at work in several cases of HIV transmission through infected blood products that could have been more thoroughly screened.[115] In one of these cases, the court wondered whether the then prevailing professional standard of care itself constituted negligence.[116]

---

[114] *Pederson v. Dumouchel*, 431 P.2d 973 (Wash. 1967).
[115] *United Blood Services v. Quintana*, 827 P.2d 509 (Colo. 1992); *Snyder v. Am. Ass'n of Blood Banks*, 676 A.2d 1036 (N.J. 1996).
[116] *Advincula v. United Blood Servicers*, 678 N.E.2d 1009 (Ill. 1996).

SUMMARY POINTS

STANDARD OF CARE

- Breach of duty of care means falling below the standard of care. In medical malpractice, an ordinarily skilled practitioner in that specialty, not one with exceptional or the highest skills, is used as the standard. The standard can usually be satisfied as long as there is a respectable minority with that view.
- An expert witness is generally needed to articulate this standard.
- In most jurisdictions, PDR warnings can be used as evidence of the standard of care, but they are not dispositive of the issue, and expert testimony is still required.
- Where negligence can be inferred, e.g., foreign bodies left within body cavities, the plaintiff can invoke the doctrine of *res ipsa loquitur*, which does not require expert testimony, to win the case.
- A defendant will usually be found negligent if there is a violation of a statute or regulation enacted to protect victims such as the plaintiff against such kind of harm.

5

Causation

After the plaintiff has established that the doctor owes a duty and experts have persuaded the court that there has been a breach of the standard of care, the plaintiff will still need to prove causation to have an actionable negligence case. This can be more complex than meets the eye. There are two types of causation, and the plaintiff must prove both. They are factual cause and proximate cause, each with its own separate requirements. Whether the defendant's conduct was a proximate cause of the plaintiff's harm is often at issue in malpractice litigation, and the plaintiff has the burden of proving proximate causation by a preponderance of evidence.

The concept of proximate causation has confused generations of students, but the basic idea is simply to show a reasonable causal connection between the negligence and the harm, i.e., that the substandard care caused the injury in a foreseeable manner. To analyze causation issues systematically, one has to separately identify factual cause issues and proximate cause issues. To make matters worse, the term 'legal cause' is sometimes used interchangeably with the term 'proximate cause.' And of course, there can be more than one proximate cause for any given injury. The California Supreme Court has rejected confusing jury instructions regarding proximate cause, and has suggested instead that the jury be directed to determine whether the defendant's conduct was a contributory factor to the plaintiff's injury.[117]

FACTUAL CAUSE

Factual cause is also known by several other terms like cause-in-fact, actual cause, or physical cause. It is established by the use of the 'but-for' test.

---

[117] *Mitchell v. Gonzales*, 819 P.2d 872 (Cal. 1991).

Causation

The typical law-school definition goes something like this: "*the defendant's conduct is a factual cause of plaintiff's injuries if plaintiff's harm would not have occurred but for defendant's conduct,*" or "*the defendant's conduct is a factual cause of plaintiff's injuries if plaintiff's harm would not have occurred without defendant's conduct.*"[118]

It is the plaintiff who must prove with a preponderence of evidence that the negligent act caused the injury. A good example is Roskin v. Rosow.[119] An internist prescribed Questran for a patient with hypercholesterolemia. Constipation resulted, and the patient was taken off medication after 7 months. However, she was again placed on Questran the following year, and the dosage was in fact increased from four to six packets per day. Colestid was subsequently substituted for Questran, and at about the same time, she received Codeine for pain. A month later, the patient experienced severe abdominal pain, and a barium enema revealed a perforated sigmoid colon. She underwent emergency surgery; the colon was found to be distended, with impacted feces 'the size of tennis balls.' Plaintiff sued internist, alleging negligence in prescribing Questran and Colestid after she had complained of constipation, and negligence in prescribing codeine which aggravated the constipation, caused fecal impaction, colon distension and perforation.

The defendant contended that plaintiff reported only mild constipation, and that the bowel was perforated during the barium enema, not from the use of the medications. Plaintiff demanded $500,000, which was then reduced to $300,000; defendant offered $100,000. The case went to trial, and the jury found for the defendant. Reason: plaintiff did not satisfy causation element.

When there are multiple causes of an injury, the courts may use the 'substantial factor' test to supplement the 'but-for' test. And in special circumstances, the law may shift the burden of proof to the defendants. In Ybarra v. Spangard, Mr. Joseph Ybarra developed complications following surgery for appendicitis. He suffered injuries to his neck, shoulder, and arm that eventually resulted in paralysis and atrophy. Six doctors and nurses were involved in the pre-, intra-, and postoperative care of the patient, but Mr. Ybarra was unable to prove which of the six was responsible for his injuries. It was a case of *res ipsa loquitur*, and the California Supreme Court

---

[118] Audiotapes on Torts by Professor Steven Finz, Sum & Substance First Year Superset ©1998.
[119] Roskin v. Rosow, (# 301356, San Mateo Cty Super. Ct., Cal. April 1987).

shifted the burden of proving the cause of injury to the defendants:

> "We merely hold that where a plaintiff receives unusual injuries while unconscious and in the course of medical treatment, all those defendants who had any control over his body or the instrumentalities which might have caused the injuries may properly be called upon to meet the inference of negligence by giving an explanation of their conduct."[120]

PROXIMATE CAUSE

Factual cause or cause-in-fact alone is just one step toward proving causation in the tort of negligence. The plaintiff also has to prove proximate cause, which is an expression of where the law draws the line in determining who will be held liable in the name of practical fairness and justice.[121] In other words, proximate cause speaks to the court's limit of how far it will go to impose liability upon the defendant if there are other intervening causative factors. In the words of the Court of Appeals of Arizona: *"A plaintiff proves proximate cause, also referred to as legal cause, by demonstrating a natural and continuous sequence of events stemming from the defendant's act or omission, unbroken by any efficient intervening cause, that produces an injury, in whole or in part, and without which the injury would not have occurred."*[122]

The key rule that governs proximate cause is foreseeability of result, i.e., whether the injury was likely rather than remote. If the defendant could not reasonably have foreseen such harm may result from his or her breach of duty of care, then the defendant escapes liability.

Sometimes, intervening factors come into play, creating uncertainty over whether there is legal causation. The concept is best understood by an example: Suppose D negligently broke the leg of a pedestrian as the result of careless driving. Unfortunately, the injury was worsened by the rescuer's negligence. Because rescuers can be foreseen to act negligently in emergency situations, the aggravation of the injury may be deemed a foreseeable result, and D remains liable to the pedestrian for both the original and the aggravated

---

[120] *Ybarra v. Spangard*, 154 P.2d 687 (Cal. 1944).
[121] *Palsgraf v. Long Island R. R. Co.*, 162 N.E. 99 (N.Y. 1928).
[122] *Barrett v. Harris*, 86 P.3d 954 (Ariz. 2004).

injury. Now suppose the accident victim suffered additional harm as the result of negligent surgery. The surgeon is, of course, liable for his or her own act of negligence, but D may well be liable for these additional surgical injuries if they were deemed to be foreseeable.

In the above examples, the rescuer and the surgeon are both foreseeable intervening factors. They are said to be concurring causes of harm and they do not free the original tortfeasor, D, from liability. In a recent Florida case, the District Court of Appeals reversed a lower court's judgment against the mother of a child whose tuberculosis meningitis was missed despite being symptomatic and having seen several different physicians. The Court held that since there were multiple doctors involved, i.e., concurring causes, the plaintiff was entitled to concurring cause jury instruction, whereas the instructions on legal cause that was given may have confused or misled the jury. The purpose of the concurring cause instruction was to negate the idea that a defendant is excused from the consequences of his or her negligence by reason of some other cause concurring in time and contributing to the same damage.[123]

On the other hand, an event may develop between the defendant's act and the plaintiff's injury that breaks the chain of causation. The law does not want to hold such a defendant liable when an unforeseeable intervening factor has led to an unforeseeable injury. The term used is superseding cause, which is defined as *"an act of a third person or other force which by its intervention prevents the actor from being liable for harm to another which his antecedent negligence is a substantial factor in bringing about."*[124]

Suppose an emergency room (ER) doctor was negligent in reading an X-ray that showed a fracture, and sent the patient home without benefit of surgical intervention. The next day, the attending physician, upon discovering the error, informed the on-call ER doctor, who was not the original doctor. This second ER doctor unfortunately failed to notify the patient. Did the second ER doctor's negligence free the first ER doctor from liability? In other words, is this a superseding cause? In a case presenting with such facts, the Sixth Circuit Court held that this was a superseding cause relieving the first doctor of liability.[125]

---

[123] *Hadley v. Terwilleger*, 873 So.2d 378 (Fl. 2004).
[124] Restatement (Second) of Torts §440.
[125] *Siggers v. Barlow*, 906 F.2d 241 (6th Cir. Ky, 1990).

Thus, an intervening cause is one that comes into active operation in producing the result <u>after</u> the negligence of the defendant. Depending on the facts, intervening causes may or may not free the defendant from liability. Superseding causes do, concurring causes do not.

LOSS OF A CHANCE

A patient may have lost the opportunity of avoiding or reducing harm because of the action or omission of the doctor. This is known as the 'loss of a chance' doctrine. In most jurisdictions, a defendant-doctor who deprives a patient of a chance of survival, even if slim, may be liable for part or all of the injuries. For example, a patient's family may claim that had the doctor taken an EKG, the patient would have been promptly hospitalized, and would have had a 25% chance of surviving the heart attack. Most jurisdictions would accept this as sufficient proximate causation, even if the lost chance did not reach a more-likely-than-not level, i.e., over 50% chance of surviving the heart attack with earlier hospitalization. Some courts would assess damages for all injuries that flowed from depriving the patient of the chance of survival, whereas other courts would apportion the damages accordingly, e.g., in this case, to only 25%.

In *Boody v. United States*, expert testimony established that the plaintiff had a 51% chance of a five-year survival had her lung cancer been timely diagnosed. The court ruled that a plaintiff could recover for the loss of any appreciable chance, not just one exceeding 50%, that resulted from a negligent act, which in this case, was failure to diagnose.[126] In another ruling, a Washington court held that the reduction in the chance of survival from 39% to 25% was enough to entrust the jury to decide on the issue of proximate causation.[127] The California Court of Appeals has framed the issue in a slightly different manner, reasoning that the negligent treatment amounted to contributory factors that led to the plaintiff's death, even if the likelihood of survival was less than 50% to start out with.[128]

A minority of jurisdictions take the position that the loss of a chance has to be more than 50% in order to constitute proximate causation.[129]

---

[126] *Boody v. United States*, 706 F.Supp.1458 (D.Kan. 1989).
[127] *Herskovits v. Group Health Co-op. of Puget Sound*, 664 P.2d 474 (Wash. 1983).
[128] *Bird v. Saenz*, 103 Cal.Rptr.2d 131 (Cal.App. 2001).
[129] *Grant v. American Nat. Red Cross*, 745 A.2d 316 (D.C.App. 2000).

PRE-EXISTING CONDITIONS

Some patients may suffer from pre-existing conditions that predispose them to greater injury, however trivial the inciting negligent act. The law will compensate the plaintiff for an aggravation of a pre-existing condition if such aggravation was caused by the defendant's negligence. Take as an example a diabetic with peripheral neuropathy and poor circulation. A negligently treated foot ulcer may cause the patient to end up with an amputation, whereas such a serious outcome would be less likely in a non-diabetic patient. In this example, the negligent doctor will be responsible for all injuries, including the amputation, though not for the pre-existing condition itself, i.e., the diabetic state.

The 'Eggshell skull rule' is the most extreme example that you take your victim as you find him. The doctrine originated from the 1901 English case of Dulieu v. White, where the plaintiff with a thin skull died from a minor accident, whereas a normal person would have suffered only a bump on the head.[130] The defendant was found liable for the patient's death. The 'Eggshell skull rule' is particularly relevant in medical malpractice cases. The negligent defendant will be held liable for the full extent of a victim's injuries, even if a pre-existing condition contributed to a more serious outcome. As an example, a minor drug allergic reaction may amount to little injury in most patients, but may lead to complications including death in an older patient already in fragile health.

HYPOTHETICAL ON CAUSATION

An example is offered below to help review and clarify the various issues of causation since it is such a difficult area of the law to grasp. The last two scenarios of the hypothetical case below deal with legal concepts of contributory negligence and damages, which are covered more fully in Chapters 6 and 7.

Suppose a surgeon was sloppy with aseptic techniques, placing his patients at risk for developing wound infections. Let's say this surgeon operates on P's leg for varicose veins. Consider the following scenarios:

- If P develops a wound infection that ultimately results in an amputation, the negligent surgeon is liable for all injuries including the amputation,

---

[130] Dulieu v. White, 2 K.B. 669 (1901).

because his negligence caused the injury, the risk being foreseeable with the lack of aseptic techniques.
- If P is a diabetic, and the wound infection fails to heal, and P ends up with an amputation, the surgeon is liable for all damages (defendant takes the plaintiff as he finds her, 'Eggshell skull rule').
- If P is involved in a traffic accident soon after the operation, injures her leg, and ends up with an amputation, the surgeon may escape liability altogether. The accident may be deemed by the court as an unforeseeable intervening cause that serves to cut off liability. That is, the accident is a superseding cause.
- If P is also careless, and fails to cleanse the wounds as instructed by the surgeon, the resulting infection and amputation could be said to have been partly P's own fault. The damages may be wiped out in jurisdictions (only a very few left) applying contributory negligence, or more likely apportioned in jurisdictions applying comparative negligence. P's own negligence played a causative role in the injury.
- If P develops an early wound infection, but misses her follow-up appointments and does not take her prescribed antibiotics, P has failed to mitigate her damages, and the quantum of recovery will be decreased proportionately (rule of avoidable consequences).

SUMMARY POINTS

CAUSATION

- For negligence to be actionable, the plaintiff must prove that the negligent act caused the injuries.

- Causation is the term used to describe the link between the defendant's negligence and the plaintiff's injuries. It has two components: factual cause and proximate (or legal) cause.

- The plaintiff carries the burden of showing factual cause by applying the 'but-for' or 'substantial factor' test, as well as proving proximate cause, which is about foreseeability.

- The defendant 'takes his victim as he finds him' (Eggshell skull rule) and may be liable for the 'loss of a chance.'

- A superseding cause, which is an unforeseeable independent force occurring after the tortfeasor's negligence, serves to free the defendant from liability.

6

Damages

In a tort action, 'damages' refers to the money awarded to the injured plaintiff. It should be distinguished from the word 'damage' which is a synonym for injuries. The primary purpose of awarding damages is to restore the person to his or her pre-injury state or to make the injured person whole, and monetary payment is the only way the law can compensate the injured in a civil action. The plaintiff must suffer actual harm, emotional and/or physical, in order to claim damages. Thus the refrain: '*no harm, no foul.*'

In the distant past, the death of either party extinguished all claims affecting both parties. This concept has since been reformed to recognize 'survival of the causes of action.' In other words, the lawsuit is now considered an inheritable property right, surviving the plaintiff's death. It should properly be called a surviving personal injury action, abbreviated to 'survival action.' Survival action should be distinguished from a wrongful death action which is brought by the survivors of the deceased person, seeking monetary compensation for losses that the negligent death had caused them.

Lost profits may sometimes be recoverable in tort action if evidence warrants recovery such as in the case of a plaintiff who is prevented from operating a business due to negligent acts of a physician. Likewise, recovery for pure mental suffering is also allowed in some jurisdictions, even if it is not accompanied by physical injury or impact. Other courts, however, look to whether the injuries were foreseeable rather than the type of injuries. In some jurisdictions, negligent infliction of emotional distress constitutes a separate tort, and is a valid cause of action.

To recover full damages, the plaintiff must take reasonable steps to mitigate his injury. This is also known as the 'rule of avoidable consequences.' It may be anticipatory such as wearing a seat belt or crash helmets, or post-damage such as seeking and abiding by continuing medical care. Note that this rule applies to an entirely different set of circumstances from the 'Eggshell skull rule' which refers to the negligent defendant's responsibility for all aggravating injuries relating to a pre-existing condition (see Chapter 5: Causation).

There are other forms of 'injuries' or 'harm.' Examples are wrongful pregnancy, wrongful birth and wrongful life. The first, also called wrongful conception, is usually associated with a failed sterilization procedure such as a vasectomy, and the plaintiff is suing for an unwanted pregnancy, even a normal one. Many courts have awarded damages for medical expenses, emotional distress and lost wages, but have been reluctant to award child rearing expenses. Wrongful birth refers to a claim by the parents of a child born with a disability or other injury as a result of negligent prenatal diagnosis or treatment. An example is the failure of proper genetic testing during pregnancy. Wrongful life is the companion lawsuit brought by the affected child. One should look to statutes and case law in any given jurisdiction in order to determine if, and to what extent, these causes of action are recognized, and if so, what compensatory damages are available. Consider the following hypothetical: A woman develops German measles during pregnancy. German measles is known to induce congenital deformities in the developing fetus. The doctor dismisses the concerns and does nothing. If the child is born with deformities, the mother may have a wrongful birth action and the child may have a claim for wrongful life.

Malpractice damages are of two sorts, compensatory damages and punitive damages:

COMPENSATORY DAMAGES

The court, i.e., a judge or jury, or the negotiated settlement determines the quantum of damages to compensate the injured plaintiff. What is fair, of course, is subject to debate. For the same injury, a high-earning plaintiff may be awarded a larger amount of compensatory damages than a low-earning plaintiff. Compensatory damages are awarded to 'make whole' the injured party. Negotiated settlements may be made in one lump sum or made in installments by using a structured periodic payment scheme. In the latter model, a portion of the lump sum is initially paid to the plaintiff and the balance is used to purchase an annuity from an insurance carrier that guarantees sum-certain periodic payments over a number of years or the entire lifetime of the plaintiff. Periodic payments are increasingly popular. For one thing, a lump sum award requires prudent investment to ensure future income. If this does not in fact occur, then lump sums do not serve the goal of the tort compensation system, which is to place the plaintiff in the position he or she would have been had it not been for the defendant's negligence. Lump sums can be 'guesstimates' at best, easily dissipated, and expensive.

Compensatory damages can be further divided into two categories:

A. Special or economic damages. These include lost wages, medical expenses, out-of-pocket expenses and other incidental expenses. They are also called pecuniary damages.
B. General or non-economic damages. These provide payments for general pain and suffering, and loss of consortium (loss of love, affection, society, sex and household services). They are also called non-pecuniary damages.

Damages for pain and suffering often contribute to 'runaway' jury verdicts, a belief shared by many in the healthcare and insurance industries. One indignant observer has written that: *"In making arguments for pain and suffering awards, both sides attempt to win the jurors' sympathies with highly emotional evidence. A blind plaintiff will receive careful instruction to come to court with his seeing-eye dog, and to dab at his eyes with a handkerchief."*[131] Tort reform proposals regularly aim to cap these awards at $250,000.[132]

Damages can sometimes reach unbelievable levels where there are high medical expenses and substantial lost wages, even if a plaintiff does not request compensation for pain and suffering. In <u>Warren v. Schecter</u>, an informed consent case where a California surgeon failed to warn of the remote risk of metabolic bone disease following gastric surgery, the jury awarded the following damages:

$140,840 in past medical care
$6,001,135 in present cash value for future medical care

---

[131] O'Connell, J. Offers that Can't be Refused: Foreclosure of Personal Injury Claims by Defendants' Prompt Tender of Claimants' Net Economic Losses. 77 N.W.U.L. Rev. 589, 591 (1982).
[132] AMA. U.S. House Passes Medical Liability Reform. Available at website www.ama-assn.org. Accessed June 6, 2004.

$262,562 in past loss of earnings
$572,376 in present cash value for lost future earnings

Judgment was entered requiring a lump sum payment of $1,824,285 and periodic payments for future care totaling $9.6 million over 34 years.[133] The plaintiff, a young woman, had developed gastrointestinal complications and multiple osteoporotic fractures following gastric surgery, and had to be fed a special diet through a feeding tube. She did not request compensation for pain and suffering, which would have been subject to California's cap of $250,000.[134]

PUNITIVE DAMAGES

Punitive or exemplary damages are awarded for conduct that merits punishment. Punitive damages may be awarded in aggravated negligence, where there is a reckless, willful or wanton disregard of the obvious risk of harm. There is usually a requirement that the egregious conduct be proven by clear and convincing evidence. The legal term that is often used to describe this type of negligence is 'gross negligence.' In one case, evidence that the defendant had prescribed an excessive number of birth control pills (over 1,000 pills within a time period when less than 200 were sufficient) with resulting liver complications was deemed sufficient to raise the issue of gross negligence.[135] And in another case, the jury awarded $125,000 in punitive damages in addition to compensatory damages because of the doctor's wanton failure to provide follow-up care of a child who developed fever and gangrenous toes following foot surgery.[136]

Punitive damages should not be confused with damages for pain and suffering. The latter belong to the category of non-economic or general compensatory damages. Punitive damages are rarely awarded in medical malpractice, whereas damages for pain and suffering are. When the evidence warrants an award of punitive damages, the amount is usually very high, as the idea is to punish the tortfeasor. Punitive damages are not a covered benefit under any professional liability insurance policy.

---

[133] *Warren v. Schecter*, 67 Cal.Rptr.2d 573 (Cal. 1997).
[134] Medical Injury Compensation Reform Act of 1975, Cal. Civ. Proc. Code §3333.2 (West, 1982).
[135] *Jackson v. Taylor*, 912 F.2d 795 (5th Cir. 1990).
[136] *Dempsey v. Phleps*, 700 So.2d 1340 (Ala. 1997).

JOINT AND SEVERAL LIABILITY

This term is used to describe the liabilities of multiple negligent defendants who have acted in concert to cause the plaintiff's injuries (multiple proximate causes). The law provides that these co-defendants, called joint tortfeasors, be jointly liable to the plaintiff for the damages, with apportionment according to the degree of fault. It is indeed the rule rather than the exception that several co-defendants are named in a malpractice lawsuit, and not infrequently, more than one party is held liable, e.g., the hospital, the attending doctor, and one or more consultants.

In some jurisdictions, each defendant can also be separately liable for the entire judgment, even if he or she is only partly to blame. This is called 'several liability,' and grew out of the argument that it is better to hold a single defendant, even if minimally at fault, responsible for the entire damages than to deprive the plaintiff of his or her just compensation. The defendant is entitled to contribution from the other tortfeasors.

This apparent unfairness of the law has led to the label 'the 1% law.'[137] Not surprisingly, it is a favorite tort reform target, and some states have modified their laws on joint and several liability.

SIZE OF AWARDS

Damages in malpractice actions are typically quite substantial. Multimillion dollar awards are often made for cases involving birth injuries and severe disabilities such as paraplegia. Most, however, do not exceed the $500,000 mark and over 90% of cases are settled out of court, thereby avoiding a trial.

Patient demands can appear unreasonable, even in apparently simple cases. A recent dental malpractice case in Hawaii illustrates this point: A patient alleged that a negligently performed tooth extraction led to recurrent maxillary sinusitis, impaired bodily immunity, and chronic Epstein-Barr virus syndrome. The plaintiff demanded $500,000 for all the injuries that allegedly

---

[137] Kendro, B. The 1% Law. *Haw Med J* 1985; 44:248.

flowed from the initial tooth extraction. The dentist refused to settle, and the case went to trial. The jury returned a verdict for the dentist-defendant.

Extended discussion and data on quantum of damages and severity of lawsuits are provided in Chapter 1 (Situation Critical, Prognosis Guarded) and Chapter 19 (What Do Malpractice Lawsuits Look Like?)

SUMMARY POINTS

DAMAGES

- Malpractice damages can be both compensatory and punitive.

- Compensatory damages are meant to return the victim to the pre-injury status. There are two types: Special damages to cover lost wages, medical expenses and other incidental expenses, and general damages for pain and suffering.

- Punitive damages are awarded to punish the tortfeasor, usually for an egregious act. They are rarely granted in medical malpractice cases.

7

Expert Testimony

In a malpractice trial, the plaintiff has to show via expert medical testimony that the defendant doctor has breached the duty of due care. Lay testimony is usually insufficient to define the standard of care. Thus, the need for an expert, or in many cases, several experts, is central to winning a malpractice lawsuit. Whereas it is relatively easy to line up experts who will testify on behalf of the doctor, it is often quite another story finding competent experts to testify against the healthcare professional. Without an expert, the plaintiff almost always loses.

Unless it is a clear case of *res ipsa loquitur* (circumstances obvious to a layperson or common knowledge; see Chapter 4: Standard of Care), the court will insist that expert testimony be presented by qualified experts. A nurse or other allied health professional will not suffice. However, a physician may testify outside of his or her specialty area, e.g., an internist with subspecialty training in infectious diseases was qualified as a plaintiff expert in a stroke case. In an interesting case alleging negligent postoperative care leading to serious brain injury, the appellate court held that the children's lay opinion that their mother suffered pain while she was in a persistently vegetative state was inadmissible evidence.[138]

The expert need not be of the same specialty as either party, but must possess the knowledge, skill, experience, training or education necessary for establishing the standard that is appropriate for the practitioner under the circumstances. For example, in litigated cases involving diabetic complications,

---

[138] *Cominsky v. Donovan*, 846 A.2d 1256 (Super. Ct. Penn. 2004).

the courts have disallowed an internist's standard for a general practitioner,[139] or an endocrinologist's standard for an internist.[140]

Most malpractice lawyers have a listing of available experts, derived from past experiences and word-of-mouth recommendations. Some plaintiff organizations have access to willing medical experts, and ads in the media and legal journals identify doctors wishing to act as experts. Attorneys generally seek experts who communicate well. How the expert is perceived by the jury is crucial, since believability of the person and therefore his or her testimony, influences the verdict. Qualifications may be what are initially assessed, but communication skills and demeanor, which include appearance, can matter more.

The going rate for an expert varies from $300–$800 per hour. These figures may be higher for depositions and testimony in open court, where the expert may charge a flat daily rate as high as $5,000 to $10,000. In Europe, expert witnesses are appointed by the courts, and are compensated according to a standard fee schedule.

In an editorial entitled "Who Should be an Expert Witness," Dr. William Parmley, then editor-in-chief of the *Journal of the American College of Cardiology*, bemoaned the state of affairs regarding experts. Some seem to be 'professional' expert witnesses who make their living mostly by testifying against other physicians. Some advertise in legal journals. Some charge exorbitant fees. Some act like 'hired guns.' The article reprinted six criteria for expert witnesses that were approved by the Board of Trustees of the American College of Cardiology in 1995. Of particular import is criterion six, which states: "*Expert witness testimony should be fair, thorough and objective. It should not exclude any relevant information that has a bearing on the case.*"[141] Various other medical associations and malpractice insurers have likewise published guidelines for those testifying as experts.[142]

---

[139] *Benison v. Silverman*, 599 N.E.2d 1101 (Ill. 1992).
[140] *Seitz v. Akron Clinic*, 557 N.E.2d 1216 (Ohio 1990).
[141] Parmley, WW. Who Should Be an Expert Witness? *J Am Coll Cardiol* 1999; 34:885.
[142] American College of Physicians: Guidelines for the Physician Expert Witness. *Ann Int Med* 1990; 113:789; Statement on the Physician Expert Witness. *Bull Am Coll Surg* 2000; 85:24–5; Guidelines for Expert Witness Testimony in Medical Malpractice Litigation. *Pediatrics* 2002; 109:974–9; The Doctor's Advocate, Third Quarter, 2003 (Publication of The Doctors Company).

Can a physician be forced to testify as an expert? In a recent Wisconsin case, the Supreme Court held that whereas a treating physician may be required to provide expert testimony regarding the care of his or her own patient, he cannot be forced to give expert testimony regarding the standard of care of another physician's patient unless the judge has determined that there are compelling circumstances. Additionally, there must be reasonable compensation and no requirement to do additional preparation in order to provide expert testimony.[143]

SCRUTINIZING THE EXPERT

The AMA considers expert medical testimony analogous to the practice of medicine, and therefore supports subjecting it to peer review. It has this to say about the ethical responsibilities of medical experts:

"Medical experts should have recent and substantive experience in the area in which they testify and should limit testimony to their sphere of medical expertise...The medical witness must not become an advocate or a partisan in the legal proceeding. The medical witness should be adequately prepared and should testify honestly and truthfully... It is unethical for a physician to accept compensation that is contingent upon the outcome of litigation."[144]

A recent AMA resolution called for the peer review of expert opinions, with disciplining of physicians who provide false statements. Loss of license has resulted in at least one case of fraudulent expert testimony.[145]

In *Austin v. America Association of Neurologial Surgeons (AANS)*, the U.S. Circuit Court of Appeals for the Seventh Circuit reaffirmed an association's right to discipline a physician for improper medical testimony. The case involved a Detroit neurosurgeon who testified for the plaintiff against a fellow association member. The injury was permanent recurrent laryngeal nerve damage following an anterior cervical fusion. The expert was suspended from AANS membership for six months, and sought damages arising

---

[143] *Glenn v. Plante*, 269 Wis.2d 575 (2004).
[144] AMA Code of Medical Ethics, 2002–3 Edition, Section 9.07.
[145] Adams, D. Physician Loses License over Expert Testimony. *AMA News* 2002; 45:10–12.

from the suspension, which reduced his expert witness earnings to 35% of his usual $220,000 per year. The court held that the neurosurgeon did not have "important economic interest" in continued membership in the association, which did not act in bad faith, and he could not recover damages for injury to his professional reputation as a result of accurate revelation of his irresponsible testimony. In supporting the AANS's position, the court wrote: *"There is a great deal of skepticism about expert evidence. It is well known that expert witnesses are often paid very handsome fees, and common sense suggests that a financial stake can influence an expert's testimony, especially when the testimony is technical and esoteric and hence difficult to refute in terms intelligible to judges and jurors. More policing of expert witnessing is required, not less."*[146]

The AMA has reported that the AANS has reviewed about 50 members for misconduct, and some 10 members have been suspended or expelled from the organization. The Florida Medical Association is willing to receive a complaint against a member for alleged false testimony, and will then forward its opinion to the Florida Board of Medicine for action.[147] In July 2002, the North Carolina Medical Board revoked the license of a Florida neurosurgeon who *"engaged in unprofessional conduct by misstating facts and the appropriate standard of care in North Carolina."*[148] And a group consisting of about 100 physicians recently formed the Coalition and Center for Ethical Medical Testimony to expose experts who falsify credentials or mislead juries. Their motto is 'Nothing but the truth.'[149]

Whereas the AMA and other medical organizations have emphasized the need for forthrightness, honesty and objectivity when providing expert medical testimony, overzealous advocates have inappropriately recommended a more aggressive and hostile approach with guidelines such as *"Be terse and dogmatic," "Don't worry about being fair to the other side," "Don't waffle or use conditional phrases,"* and *"Don't educate the opposition by citing opinion or authority that weakens or contradicts your opinions."*[150] Upon finding the many contradictions in the testimony of so-called experts, the authors of one

---

[146]*Austin v. American Assn. of Neurological Surgeons*, 253 F.3d 967 (7th Cir. Ill. 2001).
[147]AMA News, July 2, 2001. Available at amednews.com, accessed August 10, 2004.
[148]AMA News, August 19, 2002. Available at amednews.com, accessed August 10, 2004.
[149]AMA News, August 18, 2003. Available at amednews.com, accessed August 13, 2004.
[150]Fisher, CW et al. The Expert Witness: Real Issues and Suggestions. *Am J Obstet Gynecol* 1995; 172:1792–1800 (quoting instructions in package to reviewing experts from JD, MD, Inc.).

study concluded that their opinions may be "*neither reliable nor accurate for purposes of judging physician conduct.*" The authors recommended instead the use of "*independent court-appointed experts, central filing of opinion letters by experts with authoritative text citations, and a sanction process by courts and/or authorized boards for testimony that is deemed inaccurate, false, or contradictory to the standard of care.*"[151]

The courts attempt to regulate expert testimony. Federal Rule 26 requires that experts testifying in federal courts sign a written statement regarding their opinion and the basis for the opinion, their expert fees, as well as their CVs and listings of prior cases within the last four years where they provided expert medical testimony. In order to qualify as an expert, a state may require that the person devote at least 60% of professional time to clinical practice or teaching.[152] Finally, experts remain at risk for charges of perjury and/or libel if their testimony can be shown to be deliberately false.

THE *FRYE* RULE

Also known as the 'general-acceptance' principle, the *Frye* rule was enunciated by a Federal Court of Appeals in a 1923 murder case. In *Frye*, the court rejected the defendant's theory that the utterance of a falsehood requires a conscious effort which is then reflected in a greater rise in the systolic blood pressure, the so-called systolic blood pressure deception test. The court held that the test did not meet with the general recognition of the scientific community:

> "*Just when a scientific principle or discovery crosses the line between the experimental and demonstrable stages is difficult to define. Somewhere in this twilight zone the evidential force of the principle must be recognized, and while courts will go a long way in admitting expert testimony deduced from a well-recognized scientific principle or discovery, the thing from which the deduction is made must be sufficiently established to have gained general acceptance in the particular field in which it belongs.*"[153]

---

[151] Fisher, CW et al. The Expert Witness: Real Issues and Suggestions. *Am J Obstet Gynecol* 1995; 172:1792–1800.
[152] West Virginia Code §55.7B.7.
[153] *United States v. Frye*, 293 F. 1013 (D.C. Cir. 1923).

In federal courts, the Frye rule is now superseded by the Federal Rules of Evidence promulgated in 1975 and the U.S. Supreme Court's Daubert decision of 1993. However, Frye is still invoked in state courts, as exemplified by a recent New York case alleging that the erectile dysfunction drug, Viagra, caused a heart attack. The patient suffered the heart attack six days after taking Viagra, and sued Pfizer, the drug manufacturer. The Supreme Court, New York County, held that the testimony of the plaintiff expert did not meet the Frye standard for admissibility of novel scientific evidence which must be based on a principle or procedure which has gained general acceptance, though it need not be unanimously endorsed by the scientific community.[154]

On the other hand, a Pennsylvania court recently held the testimony of the defendants' expert witnesses admissible under the Frye general acceptance test. The case dealt with injuries to a woman's ureter during aorto-bifemoral bypass surgery. The plaintiff had not challenged the methodology used by the defendants' experts. Rather, it challenged the conclusions, and sought to bar the witnesses on that basis. The court ruled that the Frye test addressed the methodology, not the conclusions of experts when looking at general acceptance of the community.[155]

1975 FEDERAL RULES OF EVIDENCE (FRE) RULE 702

FRE Rule 702 was promulgated in 1975, and supplanted the Frye test by broadening the type of evidence that is admissible in court. So long as the expert is qualified, anything that will assist the jury in understanding the case will be admissible. The language of Rule 702 provides, in part:

> *"If scientific, technical, or other specialized knowledge will assist the trier of fact to understand the evidence or to determine a fact in issue, a witness qualified as an expert by knowledge, skill, experience, training, or education, may testify thereto in the form of an opinion or otherwise."*

It has been claimed that in order to satisfy the Frye test, the expert's testimony has to reach 'reasonable medical certainty' (95%) that it has achieved

---

[154] *Selig v. Pfizer*, 713 N.Y.S.2d 898 (2000).
[155] *Cummins v. Rosa*, 846 A.2d 148 (Super. Ct Penn. 2004).

general acceptance in the community. On the other hand, The Federal Rules speak in terms of medical probability, i.e., over 50% likelihood.

THE *DAUBERT* AND *KUMHO* DECISIONS

In 1993, the U.S. Supreme Court held in *Daubert* that "*'General acceptance' is not a necessary precondition to the admissibility of scientific evidence under the Federal Rules of Evidence, but the Rules ... do assign to the trial judge the task of ensuring that an expert's testimony both rests on a reliable foundation and is relevant to task at hand.*" *Daubert* was about a California lawsuit against a pharmaceutical company for birth defects allegedly caused by ingestion of the anti-nausea drug Bendectin during pregnancy. However, subjective belief or unsupported speculation was insufficient; the expert "*is permitted wide latitude to offer opinions, including those that are not based on first-hand knowledge or observation...,*" but it must have a "*reliable basis in the knowledge and experience of his discipline.*" The Supreme Court discussed several factors used to determine reliability, and they include: 1) whether the scientific information has been tested; 2) whether it has been subjected to peer review and publication; 3) recognized real or potential errors; and 4) whether it's generally accepted in the scientific community.[156] The idea is for the court to act as a gatekeeper to exclude unreliable 'junk science.' Subsequently, in *Kumho Tire v. Carmichael*, the Supreme Court extended its *Daubert* twin requirements of relevance and reliability to all expert testimony and not just scientific testimony.[157]

A federal court applied *Daubert* in ruling that the testimony of Dr. Harrington, the plaintiff's expert witness, was to be excluded because it lacked reliability. The plaintiff developed brain damage from fat embolism following hip surgery, and alleged that prolonged malleting of a hip prosthesis into the correct position was the cause of the fat embolism syndrome (FES). The court found that "*there was no evidence of widespread acceptance of Dr. Harrington's theory linking extended malleting to FES; indeed no theory linking extensive malleting to FES had ever been published.*" The court also

---

[156] *Daubert v. Merrell Dow Pharmaceurticals, Inc.,* 113 S.Ct. 2786 (1993).
[157] *Kumho Tire v. Carmichael,* 526 U.S. 137 (1999).

noted the lack of any objective source, peer review, clinical tests, establishment of an error rate or other evidence to show that Dr. Harrington followed a valid, scientific method in developing his theory.[158]

APPLICATION OF *FRYE*, FRE, *DAUBERT* AND *KUMHO*

State courts have applied these federal developments with varying consistency. Basically, the court, i.e., the judge, decides whether someone will be accepted as an expert, and it is granted great latitude in this regard. It is then up to the jury to decide on the credibility of the testimony once it is admitted into evidence. In *Larsen v. State Sav. & Loan Ass'n*, the Hawaii Supreme Court specified three decisions the trial court must make before admitting expert testimony into evidence. They are: (1) the witness is in fact an expert; (2) the subject matter of the inquiry is of such a character that only persons of skill, education, or experience in it are capable of a correct judgment as to any facts connected therewith; and (3) the expert testimony will aid the jury to understand the evidence or determine the fact in issue.[159] The Court explained that *"It is not necessary that the expert witness have the highest possible qualifications to testify about a particular matter... but the expert witness must have such skill, knowledge, or experience in the field in question as to make it appear that his opinion or inference-drawing would probably aid the trier of fact in arriving at the truth... Once the basic requisite qualifications are established, the extent of an expert's knowledge of the subject matter goes to the weight rather than the admissibility of the testimony."*

In a Hawaii malpractice case where the plaintiff alleged negligence in the treatment of a knee infection, the Court reiterated the *Larsen* ruling that *"Liberality and flexibility in evaluating qualifications should be the rule; the proposed expert should not be required to satisfy an overly narrow test of his own qualifications. The trial court has wide discretion in determining the competency of a witness as an expert with respect to a particular subject."*[160]

All in all, the guidelines issued by the Federal Rules and the Supreme Court have not been entirely successful in ensuring consistency regarding

---

[158] *Domingo v. T.K.*, 289 F.3d 600 (9th Cir., Haw. 2002).
[159] *Larsen v. State Sav. & Loan Ass'n*, 640 P.2d 286 (Haw. 1982), as cited by the Court in *Bobbitt v. Chow*, 65 P.3d 182 (Haw. 2003).
[160] *Bobbitt v. Chow*, 65 P.3d 182 (Haw. 2003).

expert medical testimony. One academician explained it this way: "*The medical and legal professions have a tradition of mutual wariness that has impeded effective cooperation in developing consistent standards for medical testimony.*"[161]

---

[161] Kassirer, JP and Cecil, JS. Inconsistency in Evidentiary Standards for Medical Testimony: Disorder in the Courts. *JAMA* 2002; 288:1382–7.

SUMMARY POINTS

EXPERT TESTIMONY

- In a malpractice action, testimony by an expert is generally required to establish the standard of care.
- Witnesses are qualified as experts by virtue of their skills, experience and education.
- Medical malpractice experts are usually physicians in the same or related specialty.
- Expert testimony does not have to meet the 'general acceptance' test (*Frye* test), but it must be relevant and reliable.
- Fair and objective expert medical testimony with consistent standards will help jurors arrive at the truth.

8

Affirmative Defenses

To win a malpractice lawsuit, the plaintiff must prove all four elements of negligence: duty, breach, causation and damages. The evidence must rise to the level of probability, else the defendant escapes liability. In practice, this concept is reduced to whether the jury finds the evidence more persuasive for one side than the other. Apart from the merits of the case, experienced trial lawyers have consistently emphasized the importance of the credibility and demeanor of the defendant and his or her experts in the courtroom. The arrogant or sloppily dressed doctor, or the know-it-all expert, may work to weaken the defendant's case.

The law allows for affirmative defenses that can defeat a malpractice action even if the evidence satisfies all the four elements of negligence. These affirmative defenses are shown in Table 5. Historically, sovereign immunity and charitable immunity were also on this list.

STATUTE OF LIMITATIONS

At common law, there was no time limit that barred a plaintiff from bringing a claim, although there was a so-called 'doctrine of latches' that foreclosed an action that had long lapsed. However, statutory changes in the law now require that complaints be brought in a timely manner so that the evidence remains fresh, accurate and reliable. Another reason is to provide repose to the wrongdoer, i.e., relieving him or her from worrying for an indefinite period of time whether a lawsuit will be brought. This time period, during which a lawsuit must be filed, is termed the statute of limitations. It is two years for the tort of negligence in most jurisdictions, although states like Tennessee and California place a one-year limit on medical malpractice claims. So does Ohio, which further states that a cause of action for

Table 5: Affirmative Defenses

1. Statute of Limitations
2. Contributory or Comparative Negligence
3. Assumption of Risk

medical malpractice accrues at the latest when the physician-patient relationship finally terminates.[162]

The statute of limitations does not necessarily start to run from the date of injury or the act of negligence, but from the time of reasonable discovery of the harm caused. Stated more formally, the period commences at the time the cause of action 'accrues,' and accrual is usually defined by statute. Thus, if there is a failure to timely diagnose and treat a cancerous condition and the patient suffers injuries several years later, suit can be filed within two years of discovering the fact that the injury was the result of such failure. In malpractice cases involving minors, the statute may be tolled, i.e., halted, until the minor reaches the age of majority. _Chaffin v. Nicosia_ dealt with such a situation. The plaintiff was born blind in one eye as the result of forceps delivery, and brought suit when he was 22-years old. Indiana had two statutes on the issue, one requiring a malpractice suit to be brought within two years of the incident, and the other allowing a minor to sue no later than two years after reaching the age of 21. The Indiana Supreme Court allowed the case to go forward, reversing the lower court's decision barring the action.[163]

Patients who are injured from malpractice may not always be aware that a negligent act had taken place, or that the injury is associated with the negligence. Recognizing this, all statutes of limitations allow accrual to begin from the date the injury is discovered, or should have been discovered if the party had exercised reasonable diligence, rather than the actual date of the negligent act or resulting injury. This is termed the discovery rule. In cases of fraudulent concealment of a right of action, the statute may be tolled during the period of concealment. Some jurisdictions incorporate an additional provision called a statute of repose that requires the filing of the action, in any event, within three years (the usual two plus one) of the negligence.

---

[162] _Kraus v. Cleveland Clinic_, 442 F.Supp. 310 (Ohio 1977), referring to R.C. Ohio §2305.11(A).
[163] _Chaffin v. Nicosia_, 310 N.E.2d 867 (Ind. 1974).

Generally speaking, courts are hostile towards attempts by the defense to use the statute of limitations to bar recovery, because this deprives the injured plaintiff of an otherwise legitimate claim. In a typical example, the defendants sought to dismiss the case (so-called motion for summary judgment) by arguing that the plaintiff filed suit some 32 months after she had developed Sheehan's syndrome following post-partum hemorrhagic shock, and this exceeded the two-year statute of limitations. The court ruled that *"Since reasonable minds could differ as to when the injury and its operative cause should have been discovered by a reasonably diligent patient, the timeliness of the plaintiff's claims should be decided by a jury and the motions for summary judgment will therefore be denied."*[164]

CONTRIBUTORY AND COMPARATIVE NEGLIGENCE

A victim may not recover if the defendant can show that the victim was also at fault. At common law, any degree of fault on the part of the plaintiff, called contributory negligence, would constitute a complete defense, and totally bar the plaintiff's action. In the earliest English case on contributory negligence, <u>Butterfield v. Forrester</u>, the defendant negligently left a pole lying across the road. The plaintiff was riding along the road and was injured as a result of the collision. The defendant escaped liability because the plaintiff could have avoided the accident if he had not been riding so fast. He was found to be contributorily negligent and recovered nothing.[165]

This rule was felt to be overly harsh to the victim who may only be slightly at fault. The law gradually changed to 'comparative negligence,' where the amount of damages is proportionately reduced by the percentage of plaintiff's contributory fault. The idea is that since the plaintiff's conduct falls below the standard of care and is a legally contributing cause, the damages ought to be apportioned. Comparative negligence currently operates in virtually all jurisdictions. The plaintiff who is 30% negligent will be able to recover 70% damages. In some states, if the plaintiff is more than 50% negligent, i.e., greater than the defendant's, then no recovery is allowed. This maxim is called modified comparative negligence. In other jurisdictions that are even more sympathetic to the tortfeasor, an equal fault on the part of the plaintiff, 50% fault, bars all recovery.

---

[164] <u>Lomeo v. Davis</u>, 53 Pa. D. & C.4th 49 (Pa. Com. Pl. Jul 24, 2001).
[165] <u>Butterfield v. Forrester</u>, 103 ER 926 (1809).

In order to counter the defense's allegation of contributory negligence, the plaintiff must show that the action taken was reasonable for his or her own welfare. In *Weil v. Seltzer*, the plaintiff was treated for many years with steroids that his doctor represented to be antihistamines. He developed many steroid complications, but continued to use the drug. His sudden death at age 54 was caused by a saddle block pulmonary embolus that contained bone marrow fragments, thought to have originated from steroid-induced osteoporotic bones. The court dismissed the defense of contributory negligence as the patient was merely taking what his doctor had prescribed, and could not have known of the negligent treatment.[166]

In a recent Tennessee case, the state Supreme Court held that *"principles of comparative fault did not apply such as to allow fault to be assessed to patient, and thus jury should not have been allowed to consider patient's antecedent negligence in assessing fault."* The patient sustained a cardiac arrest because of negligent monitoring during a CT scan. His antecedent negligence stemmed from the fact that he had alcohol in his blood upon arrival at the hospital following a car accident. The Court found the hospital 100% liable and disallowed the jury's assessment of 30% comparative fault.[167]

A related doctrine is the 'Avoidance of Consequences' which operates after a negligent act has taken place, but where the plaintiff has the opportunity to mitigate, i.e., avoid or reduce, the adverse results. An example is the diabetic's failure to properly care for a foot ulcer that had been caused by the podiatrist's negligent toe-nail clipping. In such a case, the damages will be reduced to compensate for what the injuries would have been had the plaintiff avoided the consequences. Contributory negligence, in contrast to avoidance of consequences, usually operates before or together with the tortfeasor's act.

ASSUMPTION OF RISK

If a plaintiff is fully aware of the risk to which he or she is exposed to, and voluntarily assumes that risk, then there should be no recovery of damages if harm results. This defense has been asserted most prominently in sports activities, such as boxing, where serious injuries are an integral known risk of the activity.

---

[166] *Weil v. Seltzer*, 873 F.2d 1453 (D.C.Cir. 1989).
[167] *Mercer v. Vanderbilt University, Inc.*, 134 S.W.3d 121 (Tenn. 2004).

An example of a successful assumption of risk defense in medical malpractice took place in California where a patient voluntarily and actively sought unorthodox natural herbal treatment for breast cancer after she refused conventional therapy. She received full disclosure of the nature of the experimental treatment protocol and the court therefore rejected her subsequent claim for damages. Giving informed consent to non-conventional experimental therapy that may worsen a patient's condition can be said to constitute assumption of risk.[168]

The Restatement of Torts defines assumption of risk to mean that the plaintiff fully understands the risk and nonetheless chooses voluntarily to take it.[169] The Restatement is an authoritative construct of the law by a group of respected legal scholars and is regarded as an important secondary legal authority. Case law and statutes are considered primary legal authorities.

The assumption of risk defense is called *volenti non fit injuria* in English law. It is basically a defense of consent. However, mere knowledge of risk does not necessarily imply consent. For example, a plaintiff once accepted a ride from a drunk driver and sustained injuries in a subsequent accident. The court ruled that *volenti* did not apply, unless the drunkenness was so extreme and so glaring that accepting the ride was equivalent to walking on the edge of an unfenced cliff.

In virtually all jurisdictions, consent issues in medical malpractice are now dealt with under the doctrine of informed consent rather than as an affirmative defense of assumption of risk.

---

[168] *Schneideer v. Revici*, 817 F.2d 987 (2nd Cir. N.Y. 1987).
[169] Restatement of Torts, §496-C.

SUMMARY POINTS

AFFIRMATIVE DEFENSES

- Affirmative defenses are available to the defendant even after the plaintiff has proven the elements of negligence.

- These defenses are the statute of limitations, contributory or comparative negligence, and assumption of risk (now dealt with under doctrine of informed consent).

- A lawsuit must be brought within a prescribed time interval, usually two years, which is specified by statute. The accrual of the cause of action begins after the patient has discovered or reasonably should have discovered the negligence and injury.

- Courts are generally hostile to the use of statute of limitations to dismiss an otherwise meritorious lawsuit.

- Contributory or comparative negligence is an expression of unreasonable conduct on the part of the plaintiff that contributes to the injury. Damages are then usually apportioned between the parties.

- Assumption of risk describes the knowing and voluntary acceptance by a plaintiff of the risks inherent in the activity, and serves as a total bar to recovery of damages if harm results.

9

Physician Countersuits

Many doctors believe that eager attorneys readily file malpractice lawsuits, even those entirely without merit. A recent survey by Doctors Company, a physician-owned malpractice carrier, showed that 72% of doctors view their patients as potential adversaries. Because of the contingency fee system, some suits may be filed to intimidate a doctor into settling. Additionally, lawyers are allowed extreme leeway in their advocacy for their clients, and the losing plaintiff does not have to pay for costs or attorneys' fees. This litigious environment has led to a counter-attack by the medical profession, and some have actually gone so far as to counter-sue the patient and/or the attorney after they have been exonerated in the initial lawsuit. The process is tedious, expensive, and generally unsuccessful, notwithstanding a few well publicized successes.[170]

MALICIOUS PROSECUTION

The usual legal theory that a malpractice countersuit is premised upon is malicious prosecution. Strictly speaking, this is a term that is used in criminal law, and the equivalent in civil law is 'wrongful use of civil proceedings.' The term malicious prosecution, however, has remained in use in civil actions. Courts are generally hostile to such lawsuits because public policy encourages unfettered resort to the courts for redress, and the law seeks to protect those whose claims are reasonable and are commenced in good faith.

In order to mount a successful countersuit under the legal theory of malicious prosecution, the doctor will have to prove the elements listed in Table 6.

---

[170] Gorney, M. The Joy of Countersuing Revisited. *The Doctors Advocate*, Available at www.thedoctors.com. Accessed January 16, 2004.

Table 6: Elements of Malicious Prosecution

1. Original instigation of a lawsuit against the doctor
2. Original proceedings terminated in favor of doctor, i.e., no malpractice liability found
3. Lack of probable cause
4. Malice
5. Special injuries

Occasionally, the doctor wins a malpractice countersuit. In *Williams v. Coombs*, a California case, a doctor was sued for wrongful death after the patient hanged herself following hospital admission for suicidal gestures. She was admitted to a private room instead of a special locked room. At trial, the jury found in favor of the physician. Thereafter, the physician sued the plaintiff attorney for malicious prosecution and intentional infliction of emotional distress. The court agreed that there was a lack of probable cause. It first advanced a two-point test: First, the attorney must entertain a subjective belief that the claim merits litigation; and second, that belief must satisfy an objective standard because the attorney must not prosecute a claim which a reasonable lawyer would not consider tenable. Finding that the attorney failed to meet the second prong of the test, the court stated that although probable cause is not the same as making a legal case (winning), an attorney must nonetheless refrain from an unsound and untenable claim. The attorney had relied exclusively on the allegations of his client, and he had not done much in the way of background research, found no cases on point, and sought advice from only one physician during a social encounter. The court reasoned that a *"litigant cannot be permitted to file suit based merely on a wing and a prayer, and then be retroactively justified by some serendipitous discovery so as not to be liable for malicious prosecution."*[171]

The claim for intentional infliction of emotional distress was dismissed because otherwise defamatory statements made in a judicial proceeding constituted a privileged publication.

---

[171] *Williams v. Coombs*, 224 Cal.Rptr. 865 (Cal. App. 1986).

The court's analysis of probable cause in <u>Williams</u> has since been undercut by other civil cases in California dealing with probable cause and malicious prosecution.[172]

In <u>Gentzler v. Atlee</u>, a cardiologist recommended that the patient go to a certain hospital for tests, and in a subsequent CABG procedure, the patient received contaminated blood products. Although the cardiologist did not recommend or participate in the surgery, he was sued for malpractice. After the trial court dismissed the action, he filed a countersuit against the attorney under Pennsylvania's statutory section for the wrongful use of civil proceedings. The court noted that the standard for probable cause is whether an attorney reasonably believes that a claim may be valid under existing or developing law, and that this determination is a matter of law, i.e., up to the court to decide. The court reasoned that under the facts of this case, there was no probable cause, as there was no informed consent issue, and the cardiologist did not himself order the administration of the blood products.[173]

However, most malicious prosecution actions fail. The cases below illustrate some of the legal hurdles one may encounter in such a countersuit. In <u>Dutt</u>, the attorney promptly withdrew his original malpractice lawsuit after receiving an unfavorable report from his own expert witness. A countersuit followed. The court held that as regards the malicious prosecution action, the attorney had probable cause to file the malpractice action. The court adopted an objective standard, and concluded that under the facts, a reasonable attorney would have believed that the malpractice action was tenable. The patient's condition initially deteriorated under the care of the physician, and improved after other doctors became involved in the case. The medical records corroborated the patient's story. The court stated that there was no absolute requirement to obtain an expert opinion before filing a medical malpractice suit, and also dismissed an abuse-of-process claim (see below) after finding that there was no ulterior motive than to resolve a legal dispute.[174]

It is even harder to meet the malice requirement. Allegations of willful and wanton misconduct are not synonymous with malice, particularly where no improper motive was suggested.[175] In the view of one court, the action of

---

[172] <u>Sheldon Appel Co. v. Albert & Oliker</u>, 47 Cal.3d 863 (1989).
[173] <u>Gentzler v. Atlee</u>, 443 Pa. Super. 128 (1995).
[174] <u>Dutt v. Kremp</u>, 111 Nev. 567 (1995).
[175] <u>Berlin v. Nathan</u>, 64 Ill.App.3d 940 (1st Dist. 1978).

an attorney who signed and amended a complaint without first reading the complaint did not constitute malice sufficient to support a malicious prosecution action.[176] A contingent-fee arrangement, even of large magnitude, cannot be used as evidence of improper motive or malice.[177] Neither is attorney negligence or incompetence, which one court gratuitously editorialized: "*If that constitutes malice, the courtrooms are full of malicious attorneys.*"[178] In that particular case, a surgeon was alleged to have injured a child's testicle although at trial, all expert witnesses, including two of the plaintiff's own experts, testified that there was no evidence of injury. Apparently, the attorney had not spoken to his own witnesses.

Note that for a malicious prosecution lawsuit to prevail, the plaintiff-doctor has to satisfy the 'special injury' requirement. It is not always clear what this entails, but the injury has to be beyond "*anxiety, loss of time, attorney's fees and the necessity to defend one's reputation.*"[179] In one case, the court ruled that a plaintiff in a malicious prosecution action may recover for "*humiliation, mortification and loss of reputation.*"[180] In other cases, courts have either declined to characterize as special injuries those "*normally incident to the service of process on anyone involved in a legal suit*" or held that "*plaintiffs' claimed damage to his professional reputation does not constitute special injury.*"[181]

Not all jurisdictions require the special injury element. The Illinois Code of Civil Procedure §2-109, for example, specifically grants this exemption. ("*In all cases alleging malicious prosecution . . . the plaintiff need not plead or prove special injury. . . .*") However, in exchange, Illinois disallows exemplary or punitive damages. The statute was passed in the midst of malpractice crisis of the 80s. In a challenge to the constitutionality of the statute, the Illinois court ruled that the state interest was a legitimate one, being to ease "*the burden of bringing a malicious prosecution action for healthcare professionals with the specific intent of not only discouraging the filing of frivolous medical*

---

[176] *Raine v. Drasin*, 621 S.W.2d 895 (Ky. 1981).
[177] *Miskew v. Hess*, 21 Kan.App.2d 927 (1996).
[178] *Spencer v. Burglass*, 337 So.2d 596 (La. App. 1976).
[179] *Stopka v. Lesser*, 82 Ill. App.3d 323 (1st Dist. 1980).
[180] *Raine v. Drasin*, 621 S.W.2d 895 (Ky. 1981).
[181] *Morowitz v. Marvel*, 423 A.2d 196 (D.C. 1980); *Ackerman v. Lagano*, 412 A.2d 1954 (N.J. 1979).

malpractice lawsuits, but also as a way of punishing those plaintiffs who bring baseless medical malpractice claims."[182]

In a twist to the usual claim that the doctor was defamed as the result of a frivolous malpractice suit, Holloway v. Texas Medical Association was brought by a malpractice attorney against the doctor and the Texas Medical Association for libel, slander and civil conspiracy arising out of a million-dollar countersuit that was brought against him. It could be called a counter counter-suit. The Court of Appeals of Texas ruled that statements made by the physician with regard to the countersuit against the malpractice attorney fell within the privileged exception to defamation, and that the Texas Medical Association's financial contribution to countersuit actions was a protected mode of expression.[183]

ABUSE OF PROCESS

Another legal theory for a malpractice countersuit is abuse of process. Here, the doctor is saying that the plaintiff attorney intentionally used the legal system for a purpose other than that for which the process was designed. Neither a showing of lack of probable cause nor favorable termination of the original lawsuit is necessary. The doctor may counterclaim for abuse of process in the same action without awaiting favorable resolution of the initial lawsuit.

Here too, the doctor's victory is rare. However, in Bull v. McCuskey, a physician was successful in his suit under this theory. He asserted that the attorney had filed a malpractice suit with the motive of coercing a nuisance settlement (attorney offered to settle for $750). The case involved an elderly woman who sustained fractures following an auto accident. She developed bedsores after refusing to follow instructions rather than because of lack of care by the hospital staff. The attorney did not examine the medical records, conferred with no physician, retained no expert, and took no depositions. At trial, which was won by the physician, the attorney called the physician

---

[182] Miller v. Rosenberg, 749 N.E.2d 946 (Ill. 2001).
[183] Holloway v. Texas Medical Association, 757 S.W.2d 810 (Tex. 1988).

incompetent, a liar and a scoundrel. In the abuse of process action that followed, the physician won a jury verdict for $35,000 in compensatory damages, and $50,000 in punitive damages, which was upheld on appeal.[184]

OTHER CAUSES OF ACTION

Creative doctors and their lawyers have advanced numerous other legal theories in their counter offensives against plaintiff attorneys. Among these are infliction of emotional distress, negligence, defamation, invasion of privacy and the tort of outrage. The claim of professional negligence basically asserts that the attorney has a duty to the physician to not bring a frivolous malpractice suit. It has been put more colorfully: *"The school bully who harasses innocent children in the playground isn't stopped by reporting to his parents [the bar association or licensing authority] — he's only stopped by a punch in the face. Lawyers who abuse the court system similarly can only be stopped through the same system."*[185] These countersuits, however, are rarely successful. Defamation suits usually fail because of the privilege accorded attorneys regarding statements made during the judicial process.

In a New York case, a physician was initially sued for malpractice, although he did not participate in the care of the patient during the course of the illness. He countersued the attorney for $200,000 for the latter's reckless disregard for the truth of the allegation. The court, however, concluded that *"to hold an attorney personally responsible for instituting a frivolous action on behalf of a client would operate to discourage free resort to the courts for the resolution of controversies, contrary to public policy."*[186] In another case, a Texan court refused to entertain an action unless based on malicious prosecution, abuse of process, contempt of court or invasion of privacy.[187]

FEDERAL RULE 11

A better way to deal with frivolous suits is to invoke the so-called Rule 11 sanction. The U.S. federal courts in 1983 adopted a tougher standard for

---

[184] *Bull v. McCuskey*, 96 Nev. 706 (1980).
[185] Annas, G. Doctors Sue lawyers: Malpractice Inside Out. Hastings Center Report, October 1977, pp. 15–6, quoting Dr. Leonard Berlin, a recent victor in a malpractice countersuit.
[186] *Drago v. Buonagurio*, 391 N.Y.S.2d 61 (Sup. Ct. Schenectady Co. 1977).
[187] *Wolfe v. Arroyo*, 543 S.W.2d 11 (Tex. Ct. App. 1976).

attorneys regarding the filing of a lawsuit. Under Federal Rule 11 of the amended Rules of Civil Procedure, "*a signature of an attorney or party constitutes a certificate by him that he has read the pleading motion, or other paper; that to the best of his knowledge, information, and belief formed after reasonable inquiry it is well grounded in fact... and that it is not interposed for any improper purpose, such as to harass or cause unnecessary delay or needless increase in the cost of litigation.*"

Rule 11 requires the judge to impose sanctions on the party in violation of the rule. Attorneys are required under Rule 11 to file an action only after reasonable inquiry of the facts and law at issue.

An example of a Rule 11 sanction involved a suit that was filed by two chiropractors in Colorado that alleged antitrust violation by a medical facility that denied them hospital admitting privileges. The facts revealed that the chiropractors had in fact never applied for those privileges. The judge imposed sanctions of $38,500.[188]

Rule 11 is a federal rule of court, i.e., applicable only in federal courts, but many state courts have adopted a similar procedural safeguard against the filing of frivolous suits.

WHAT THE FUTURE HOLDS[189]

Despite the difficulties in winning malpractice countersuits, they are unlikely to go away. Insurance carriers are not routinely supportive of such actions, and will usually not underwrite the costs for such a lawsuit. Instead, state medical associations and other independent groups have taken up the fight. The latest arrival on the scene is Medical Justice Corp., based in Greensboro, N. Carolina, which directly warns the plaintiff that a countersuit will be filed if the malpractice lawsuit is found to be frivolous. It will also investigate expert witnesses if there is reason to believe they have committed perjury or given false testimony. Medical Justice covers about 500 members in thirty states, and each member pays between $625 and $1,800 per year. The group is reportedly about to file its first countersuit.

---

[188] *Colorado Chiropractic Council v. Porter Memorial Hospital*, 650 F. Supp. 231 (Co. 1986).
[189] Doctors Take the Offensive. *The Wall Street Journal*, March 23, 2004, p. D3.

Some have attempted to go even further. An Internet-based database company named DoctorsKnow.Us was to provide listings of litigious patients, trial attorneys and expert witnesses known to be involved in medical malpractice lawsuits. However, its services were terminated after patient advocacy groups registered their vigorous opposition, claiming that the service threatened potential patients from being able to assess needed medical care.

SUMMARY POINTS

PHYSICIAN COUNTERSUITS

- Physician countersuits have only rarely been successful in seeking redress against a frivolous malpractice lawsuit.

- These lawsuits are usually premised on two legal theories: malicious prosecution, and abuse of process.

- The key elements of malicious prosecution include exoneration in the initial malpractice lawsuit, lack of probable cause and presence of malice.

- Abuse of process speaks to using the legal system for an illegitimate purpose. Its use does not necessitate winning the underlying malpractice lawsuit.

- Rule 11 is a federal rule that requires the judge to impose sanctions where there has been a failure to conduct 'reasonable inquiry' prior to filing the lawsuit.

10

Informed Consent

Human experimentation and end-of-life decisions to forgo life-sustaining treatments are well-understood examples where specific consent from the patient is necessary. The law today requires that the patient or legally authorized representative be the sole and ultimate arbiter to accept or reject all forms of treatment, not just those associated with clinical research or end-of-life care. There was a time when the healthcare provider's recommendation was basically thrust upon the patient, i.e., 'doctor knows best,' or so-called paternalistic model. However, the present-day standard is that absent consent, either explicit or implied, the doctor has no authority to examine or treat the patient. This consent, to be valid, is obtained after the patient has been given the opportunity to make a free and informed decision. In other words, informed consent constitutes the legal authority to treat and informed refusal is a denial of such authority. It is a pervasive principle evident in clinical situations ranging from giving permission for surgery to signing out against medical advice. Generally speaking, the physician is obligated to follow the patient's wishes as long as the patient has decision-making capacity.

The law of battery renders one liable if one acts intentionally to inflict an offensive or harmful touching of another person without consent. Thus, medical treatment without the patient's consent has been characterized as an assault and battery in some jurisdictions. This had its genesis in a 1914 New York case in which a fibroid tumor was removed from a woman while she was under ether anesthesia, despite the fact that she had specifically refused the operation. The facts of the case prompted the famous Judge Cardozo to write:

> "Every person of adult years and sound mind has a right to determine what shall be done with his own body; and a surgeon who

performs an operation without his patient's consent commits an assault, for which he is liable in damages."[190]

Without consent, except under rare circumstances, the physician has no authority to treat, even if the withholding of beneficial treatment is expected to lead to patient harm. The best known example is the refusal of blood transfusions by Jehovah's Witnesses. Courts have generally ruled that competent adults can refuse blood even if death would result; but where a third-party interest is at stake, e.g., a fetus at term, the healthcare provider may order that blood be administered.[191]

Today, most states consider treatment without informed consent to be covered by the tort of negligence rather than the intentional tort of battery. This has important legal implications because the two torts have different statutes of limitations. In addition, an action in battery requires neither expert medical testimony nor the showing of specific damages. In a dramatic case where a battery claim was brought, the defendant dentist had extracted all thirty-two teeth during a single visit without the patient's full understanding of and consent to the procedure.[192]

WHAT CONSTITUTES INFORMED CONSENT?

In order for patients to meaningfully give their consent to treatment, they must have sufficient information regarding the doctor's treatment plans. The consent must also be given voluntarily, and not be a coerced one. The notion of patient autonomy is so entrenched that all states have enacted legislation on the issue. The law imposes upon the practitioner the duty to disclose three fundamental aspects of treatment, easily remembered by the mnemonic **PAR:**

P = Procedure

A = Alternatives

R = Risks

---

[190] *Schloendorff v. Society of New York Hospital*, 101 N.E. 92 (N.Y. 1914).
[191] *Raleigh Fitkin-Paul Morgan Memorial Hospital v. Anderson*, 201 A.2d 537 (N.J. 1964).
[192] *Blanchard v. Kellum*, 975 S.W.2d 522 (Tenn. 1998).

Procedure: The doctor should explain the diagnosis and the contemplated procedure or treatment plan and expected results.

Alternatives: The patient should be informed of all alternatives, including that of doing nothing.

Risks: The material risks should be discussed with the patient, including the risk of refusal of treatment. What constitutes a material risk is at the heart of the controversy surrounding the informed consent doctrine. Generally speaking, the patient should learn of all serious risks, even if unusual or rare.

Hawaii's law, which is typical of informed consent statutes in general, is codified in Hawaii Revised Statutes §671-3. Recently amended by the 2003 legislature, the risk disclosure portion of the statute is set out in (b) 5:

> *(b) The following information shall be supplied to the patient or the patient's guardian or legal surrogate prior to obtaining consent to a proposed medical or surgical treatment or a diagnostic or therapeutic procedure:*
>
> *5. The recognized material risks of serious complications or mortality associated with:*
>   *A. The proposed treatment or procedure;*
>   *B. The recognized alternative treatments or procedures; and*
>   *C. Not undergoing any treatment or procedure ...*

Note that Hawaii law now requires the discussion of "*the recognized material risks of serious complications or mortality,*" whereas the law it replaced used the phrase "*the recognized, serious, possible risks, complications and anticipated benefits.*" This is thought to clarify the doctor's duty regarding informed consent, but in reality, the new language is unlikely to have a significant practical effect. An earlier 1976 version of the law more reasonably required the disclosure of "*probable risks and effects.*"

'MATERIAL RISKS'

What objectively constitutes a material risk that needs disclosure continues to plague doctors. The likelihood and severity of a risk are obvious factors to consider. However, in one case, a 1% risk of hearing loss required disclosure,[193] whereas in another, the court appeared to say that a 1.5% chance of visual loss did not.[194]

Warren v. Schecter is probably one of the most dramatic cases to ever confront the material risk issue. The plaintiff won a $9.6 million judgment for failure of the doctor to disclose risk of osteoporosis.[195] Dr. Schecter had performed gastric surgery on Janet Warren for peptic ulcer disease, and had warned the patient of the risks of bowel obstruction, dumping syndrome, and anesthetic death. He did not believe osteoporosis, osteomalacia and bone pain were risks of surgery, and so did not discuss these risks with her. Warren testified at trial that had Dr. Schecter warned of the risk of metabolic bone disease, she would not have consented to the surgery. A second surgery was undertaken because Warren developed post-op dumping syndrome and alkaline reflux gastritis, and the surgeon again failed to advise her of the risk of metabolic bone disease. She again testified that she would not have consented to the second surgery had she been duly advised of the risk.

The plaintiff subsequently developed severe osteoporotic fractures, and filed a malpractice lawsuit alleging that Dr. Schecter was liable under an informed consent theory for performing surgery without advising her of the risk of bone disease. The jury found that Dr. Schecter did not disclose to Warren all relevant information which would have enabled her to make an informed decision regarding surgery and that a reasonably prudent person in Warren's position would not have consented to surgery if adequately informed of all the significant perils. Judgment was entered requiring a lump sum payment of $1,824,285 and periodic payments for future care totaling $9.6 million over 34 years.

In addition to requiring the disclosure of material risks associated with a surgical operation or a medication, courts have also looked at other aspects of disclosure in the doctor-patient relationship. These include disclosure of limited experience of a neurosurgeon in a risky procedure[196]; financial incentives

---

[193] *Scott v. Wilson*, 396 S.W.2d 532 (Tex. Civ. App. 1965).
[194] *Yeats v. Harms*, 393 P.2d 982 (Kan. 1964).
[195] *Warren v. Schecter*, 67 Cal.Rptr.2d 573 (Cal. 1997).
[196] *Johnson v. Kokemoor*, 545 N.W.2d 495 (Wis. 1996).

amounting to a breach of fiduciary responsibility[197]; and a surgeon's positive HIV status[198] or his alcoholism.[199] However, in <u>Arato v. Avedon</u>, the California Supreme Court held that the law did not require physicians to inform their terminally-ill patients of their prognosis and life expectancy.[200]

OBTAINING CONSENT

Explicit or expressed consent can be verbal or written. But consent can also be implied, e.g., stretching out one's arm for venipuncture is an act that implies consent for blood drawing. Although implied consent may be sufficient for minor procedures, written consent is recommended for surgery, invasive treatments or procedures, medical research, and release of medical information.

The doctor, not the office assistant or clinic nurse, should be the person who explains the procedures, alternatives and risks to the patient or legally authorized representative. The permission to treat is then documented in the chart. This rule is frequently violated, and the important task of obtaining consent is improperly delegated to one who may not have all the relevant medical facts. To be sure, the assistant can reinforce the information and witness the signature, but the primary task rests with the doctor. Moreover, consent for a procedure or an operation that is to be performed by one doctor cannot simply be transferred to another doctor unless there is prior authorization.

Whether the primary referring physician is also obligated to warn of material risks, or whether this is the exclusive duty of the consultant who will be performing the procedure, is a current contentious issue. This may seem an odd inquiry, as traditionally it is the consultant who obtains the consent for the intended procedure. However, where there is ongoing retention of control or supervision of the patient by the primary referring healthcare provider, some jurisdictions, including Hawaii, have held that a duty to warn also attaches to that primary provider.[201] New York also subscribes to this view, although in

---

[197] <u>Moore v. The Regents of the University of California</u>, 793 P.2d 479 (Cal. 1990).
[198] <u>Estate of Behringer v. The Medical Center at Princeton</u>, 192 A.2d 1251 (N.J. Super. 1991).
[199] <u>Hidding v. Williams</u>, 578 So.2d 1192 (La.App. 1991).
[200] <u>Arato v. Avedon</u>, 858 P.2d 598 (Cal. 1993).
[201] <u>O'Neal v. Hammer</u>, 953 P.2d 561 (Haw. 1998).

one instance, the court expressed that it was *"not necessary that every physician or health worker who becomes involved with a patient obtain informed consent to every medical procedure to which the patient submits"* and that referring physicians who do not participate in the procedure cannot be held liable for the surgeon's failure to obtain informed consent.[202]

INFORMED CONSENT AND CONSENT FORMS

Typically, the healthcare provider uses a standardized consent form with the details in fine print. This is particularly likely for consent forms that cover surgical procedures. Do not confuse informed consent, which is a process, with a signed consent form. The former is a legal principle governing treatment, the latter is merely written evidence that informed consent was purportedly obtained. The signed form can be challenged and invalidated upon a showing that it was improperly executed, e.g., a patient signing the form after the procedure was done, or evidence that the patient was never told of certain risks that were pre-printed on the form. In a Hawaii case on this matter, the court clearly showed its hostility to the use of such forms:

"Moreover, a physician may not fulfill his affirmative duty of timely and adequate disclosure by merely having the patient sign a printed informed consent form. A signed consent form is not a substitute for the required disclosure by a physician .... There is growing reason for concern that consent forms are becoming substitutes for, rather than documentary evidence of, an ongoing process of disclosure, discussion, and decision-making between physician and patient. If physicians come to believe (often erroneously) that their obligation to obtain a patient's informed consent can be satisfied by securing a signed signature — even that of a drowsy, drugged, or confused patient on an abstruse, jargon-ridden, and largely unintelligible preprinted consent form — the law's reliance on written documentation may come to pervert its central purpose in requiring informed consent."[203]

---

[202] *Nisenholtz v. Mount Sinai Hospital*, 483 N.Y.S.2d 568 (1984).
[203] *Keomaka v. Zakaib*, 811 P.2d 478 (Haw. 1991).

CAPACITY TO CONSENT

All adults of sound mind are presumed to be competent to give informed consent. The better term is 'capacity for medical decision-making.' Consent is obtained from the patient with capacity; otherwise a legally authorized surrogate decision-maker, usually the next of kin or legal guardian gives the consent. In most instances, one can determine capacity without much difficulty. A simple approach in deciding whether the patient has medical decision-making capacity is to ask two questions: (1) whether the patient understands the procedure or treatment; and more importantly, (2) whether the patient understands the consequences of accepting or refusing such treatment and alternatives. It is the attending physician's clinical duty to determine the patient's decision-making capacity. In difficult cases, a consultation with a psychiatrist may be of some help, but this is not necessary or recommended in the usual case.

For a minor, the doctor must obtain permission from the parents or guardian unless the minor is emancipated, i.e., conducting himself or herself as an adult and no longer under the control of the parents. If a relative rather than the parents accompanies the child, parental consent should still be sought. In the case of separated or divorced parents, the party having custody of the child usually gives the consent.

Special rules apply in reproductive medicine. For example, the law does not require parental consent for a minor's request for an abortion.[204] In some jurisdictions, the law may even disallow informing the parents of such a request, or it may permit parental notice, but deny them any decision-making authority.[205] Family planning, pregnancy services, and treatment for venereal diseases in minors are exempt from the requirement of parental consent.

In patients who have prepared an advance medical directive (AMD), physicians are obligated by law to follow the patient's wishes regarding life-sustaining treatment, typically at the end of life. The AMD, sometimes called a 'living will,' is drawn up when the individual is fully cognitive, and springs into effect when the person loses the capacity to decide. The patient typically identifies a surrogate decision-maker in the AMD, and this person is said to have durable power of attorney for healthcare decisions. If a healthcare

---
[204] *Planned Parenthood of Central Mo. v. Danforth*, 428 U.S. 52 (1976).
[205] *H.L. v. Matheson*, 450 U.S. 398 (1981).

provider is unable to effectuate the patient or surrogate's wishes, he or she is legally obligated to transfer the patient to another provider.

EXCEPTIONS TO INFORMED CONSENT

Under rare circumstances, informed consent may be neither possible nor necessary. Statutory provisions that protect public health and safety may mandate quarantine, examination, treatment of a patient, or referral of a death to a coroner. The following are additional legitimate exceptions to the informed consent requirement:

Emergencies: The guiding principle is whether delay in treatment in order to obtain consent would result in harm to the patient. The procedure need not be life saving, so long as the potential harm to the patient is significant. This exception is typically provided for in state statutes on informed consent, e.g., *"Nothing in this section shall require informed consent from a patient or a patient's guardian when emergency treatment or emergency surgical procedure is rendered by a healthcare provider and the obtaining of consent is not reasonably feasible under the circumstances without adversely affecting the condition of the patient's health."*[206]

Unanticipated Conditions during Surgery: This is a narrowly construed exception and comes into play when a surgeon encounters an unanticipated abnormality in the field of surgery. It is called the 'extension doctrine,' and assumes that the surgeon is using reasonable judgment. Thus, a surgeon incurred no liability for draining some ovarian cysts during the course of an appendectomy.[207] But where the surgeon operated on the left ear despite consent only for the right ear, the court held his conduct actionable as the situation was not a true emergency.[208] The condition must be one that was unforeseen, and the patient had not expressly refused such intervention. Most informed consent forms now incorporate an 'unanticipated condition' clause.

Therapeutic Privilege: If a doctor believes that the patient's emotional and physical condition could be adversely affected by full disclosure of the treatment risks, then disclosure may be legally withheld. This principle is called

---

[206] Hawaii Revised Statutes §671-3 (d).
[207] *Kennedy v. Parrott*, 90 S.E.2d 754 (N.C. 1956).
[208] *Mohr v. Williams*, 104 N.W. 12 (Minn. 1905).

'therapeutic privilege.' Courts pay lip service to this doctrine but strongly disfavor actual reliance on it. It was clearly enunciated in *Nishi v. Hartwell*, Hawaii's first case on informed consent. The plaintiff-dentist, Dr. Nishi, sought damages for below-waist paralysis following thoracic aortography. This procedure related risk was never discussed with him, purportedly because of his serious underlying cardiac status and extreme apprehension over his condition.

The case reached the Hawaii Supreme Court in 1970. The Court first defined informed consent as imposing upon a physician *"a duty to disclose to his patient all relevant information concerning a proposed treatment, including the collateral hazards attendant thereto, so that the patient's consent to the treatment would be an intelligent one based on complete information."* The court ruled that cases involving an alleged lack of informed consent *"are deemed to sound in negligence (rather than battery), as raising the question of a neglect of duty required to be observed by a physician in his relationship with his patient."*

The Court then addressed the therapeutic privilege defense that was raised by the defendant. Speaking in approving terms that betray a clear deference to the medical profession, the Hawaii Supreme Court held that: *"the doctrine recognizes that the primary duty of a physician is to do what is best for his patient, and that a physician may withhold disclosure of information regarding any untoward consequences of a treatment where full disclosure will be detrimental to the patient's total care and best interest."* Finally, the Court also held that *"the duty of a physician to make full disclosure is one that arises from the physician-patient relationship. It is owed to the patient himself, and not to his spouse or any other member of his family."*[209]

In the well-known case of *Canterbury v. Spence*, the U.S. Court of Appeals in the District of Columbia also articulated the therapeutic privilege exception to informed consent, in order to enable the doctor to withhold risk information if such disclosure would pose a serious threat of psychological detriment to the patient. However, the physician is still required to disclose that information which will not prove harmful to the patient.[210]

---

[209] *Nishi v. Hartwell*, 473 P.2d 116 (Haw. 1970).
[210] *Canterbury v. Spence*, 464 F.2d 772 (D.C.Cir. 1972).

Waiver or Risks Known to the Patient: Some patients expressly indicate that they do not wish to be informed of the treatment procedure and associated risks. This constitutes a waiver and is recognized as a legitimate exception. Waivers should be documented in writing to protect the doctor. The healthcare provider is also not obligated to disclose risks that are commonly understood, obvious, or already known to the patient.

Informed Consent not Feasible: The United States government was alleged to have used investigational drugs on military personnel during the Gulf War without their consent. In <u>Doe v. Sullivan</u>, a federal court refused to enforce the informed consent requirement because of the impracticality of obtaining consent under the circumstances.[211] This exception to the informed consent is obviously a very narrow one.

WHAT IF THE PATIENT DOES NOT ASK?

There is a misconception of the law that if the patient does not ask, then there is no duty to disclose. The doctor has an affirmative duty to disclose the material risks, irrespective of whether the patient has asked the proper questions. This was brought home decisively in <u>Keomaka v. Zakaib</u>:

> *"[Because of] the superior knowledge of the doctor with his expertise in medical matters and the generally limited ability of the patient to ascertain the existence of certain risks and dangers that inhere in certain medical treatments, it would be unfair and illogical to impose on the patient the duty of inquiry or other affirmative duty with respect to informed consent. Thus, where a patient has no duty in the informed consent context, we cannot see how the patient can be contributorily negligent... contributory negligence has no place in an action for failure to obtain informed consent."*[212]

COMMON MYTHS REGARDING INFORMED CONSENT

Myths abound regarding the doctrine of informed consent, and many physicians remain mistrustful of the doctrine. In truth, however, informed

---
[211] *Doe v. Sullivan*, 938 F.2d 1370 (D.C.Cir. 1991).
[212] *Keomaka v. Zakaib*, 811 P.2d 478 (Haw. 1991).

consent has been described as a shared process for decision-making by both patient and doctor, with the doctor providing the medical facts, and the patient exercising autonomy in accepting the recommended treatment or choosing from available options. It is improper for doctors to merely offer the various choices without expressing their opinions and recommendations.

A recent article discusses eight myths regarding informed consent. These myths should be self-evident, and serve as a useful reminder of what informed consent is not:

1. A signed consent form is informed consent.
2. Informed consent is a medical Miranda warning.
3. Informed consent requires that physicians operate a medical cafeteria.
4. Patients must be told everything about treatment.
5. Patients need full disclosure about treatment only if they consent.
6. Patients cannot give informed consent because they cannot understand complex medical information.
7. Patients must be given information whether they want it or not.
8. Information may be withheld if it will cause the patient to refuse treatment.[213]

LEGAL STANDARDS

Consider a patient who is angry over a perceived treatment mishap. In filing a suit against the doctor, the patient invariably alleges that no warning of the risks had been disclosed. Further, had the patient been warned, he or she would not have consented to the procedure and therefore would have avoided the subsequent injury.

By what measure do we decide whether the doctor had been negligent on matters relating to informed consent? The traditional rule regarding malpractice is to judge the conduct of the defendant by that expected of his peers, i.e., what is ordinarily expected of a similarly situated doctor in the same specialty (see Chapter 4: Standard of Care). However the doctrine of informed consent, although based on negligence law, has begun moving away from this long accepted position. The standard governing the disclosure of risks

---

[213] Miesel, A and Kuczewski, M. Legal and Ethical Myths about Informed Consent. *Arch Int Med* 1996; 156:2521–6.

and causation differs significantly among the jurisdictions. The traditional law employs the physician-oriented standard (also known as the professional or medical standard); however, about half the jurisdictions have abandoned this for the patient-oriented or reasonable person's standard, which is a much higher hurdle for the profession to meet. These developments have major significance for the practicing doctor, and are discussed in some detail below. Case law over the course of three decades of litigation in Hawaii is used to illustrate the evolution of the legal standard of disclosure.

The Traditional Law Governing Disclosure: The traditional standard is the physician-oriented standard — what the community of doctors would disclose. This is the legal standard for negligence actions, and the tort of the lack of informed consent, initially considered an assault and battery action, had been determined by many courts to be better analyzed under a negligence theory. Hawaii's first case of informed consent in 1970, that of *Nishi v. Hartwell* (see above, under therapeutic privilege, for facts of the case), clearly subscribed to this change, but preserved the physician-oriented standard for deciding liability:

> *"In determining the question of physician's liability for nondisclosure courts generally follow the rule applicable to medical malpractice actions predicated on alleged negligence in treatment which requires the question of negligence to be decided by reference to relevant medical standards and imposes on the plaintiff the burden of proving the applicable standard by expert testimony."*[214]

The 'New' Law: In 1972, the U.S. Court of Appeals in the District of Columbia decided an important case, *Canterbury v. Spence*, which started the movement away from the physician-oriented standard, although many jurisdictions are yet to become converts. In *Canterbury*, a 19-year-old young man developed paraplegia the day after spinal surgery. The risk of this happening was 1%, but it was not disclosed to the patient prior to surgery. The Court first reaffirmed the notion that every competent adult has a right to determine what shall be done with his own body, that informed consent entails the knowledgeable evaluation of options and risks, and that the doctor must therefore

---

[214] *Nishi v. Hartwell*, 473 P.2d 116 (Haw. 1970).

disclose all material risks. Declaring "*the patient's right of self-decision shapes the boundaries of the duty to reveal,*" the Court went on to state:

> "*A risk is... material when a reasonable person, in what the physician knows or should know to be the patient's position, would be likely to attach significance to the risk or cluster of risks in deciding whether or not to forgo the proposed therapy... any definition of scope in terms purely of a professional standard is at odds with the patient's prerogative to decide on projected therapy himself..."*[215]

With these words, the standard was heightened from what doctors ordinarily would reveal to what a reasonable person in the patient's position would want to hear. This District of Columbia decision was quickly adopted by California in Cobbs v. Grant.[216] In Hawaii, the reasonable patient standard for the disclosure of risks was first raised (under footnote 10) in the 1985 case of Leyson v. Steuermann,[217] and definitively articulated in 1995 in Carr v. Strode (failed vasectomy operation that resulted in the unplanned birth of a healthy child, where the plaintiffs alleged absence of informed consent for the vasectomy). In its decision, the Hawaii Supreme Court reversed its own 1970 Nishi ruling:

> "*We believe that the patient-oriented standard of disclosure better respects the patient's right of self-determination and affixes the focus of inquiry regarding the standard of disclosure on the motivating force and purpose of the doctrine of informed consent... aiding the individual patient in making an important decision regarding medical care. It also protects against the pitfalls of proof associated with the physician-oriented standard... Moreover, not only should the patient's decision remain at the forefront when assessing the physician's disclosure to his or her patient in each case, but we also believe that, barring situations where the therapeutic privilege exception to the physician's duty to disclose is applicable, what the medical community*

---

[215] Canterbury v. Spence, 464 F.2d 772 (D.C.Cir. 1972).
[216] Cobbs v. Grant, 502 P.2d 1 (Cal. 1972).
[217] Leyson v. Steuermann, 705 P.2d 37 (Haw. 1985).

believes the patient needs to hear in order for the patient to make an informed decision is insufficient, without more, to resolve the question of what an individual patient reasonably needs to hear in order for that patient to make an informed and intelligent choice regarding the proposed treatment... We therefore expressly adopt the patient-oriented standard applicable to a physician's duty to disclose risk information prior to treatment, and, to the extent that Nishi may be construed otherwise, it is overruled."[218]

In contrast to D.C. and states like California, Hawaii, New Jersey and Louisiana, the majority of jurisdictions still apply the physician-oriented standard to judge informed consent liability. However, an increasing number of jurisdictions are opting for the patient-oriented standard.

Causation: Because informed consent issues are covered under the tort of negligence, the plaintiff is required to prove proximate causation, i.e., that the negligent failure to disclose caused the harm. The legal question governing causation is this: If the material risks were disclosed, would the patient still have undergone the procedure, thereby incurring the same harm? And from whose viewpoint should this question be answered? In 1985, the Hawaii Intermediate Court of Appeals took its first stab at this legal issue in Leyson. The Court asked, somewhat rhetorically, "*whether this question* [of causation] *should be examined from the viewpoint of Leyson* [the patient who was injured from the long-term use of steroids for psoriasis], *a reasonable person in Leyson's position, an ordinary person in Leyson's position, or some other viewpoint.*" The Court concluded that the "*reasonableness factor would weigh heavily, if not predominantly, in the determination of the credibility and weight of the evidence on the subject,*" and opted for "*the application of a modified objective standard that determines the question from the viewpoint of the actual patient acting rationally and reasonably.*"

This so-called 'modified objective standard' (really a subjective standard) for deciding causation was clarified in 1995 when the Hawaii Supreme Court decided Bernard v. Char, a case involving a patient who elected to undergo a tooth extraction rather than a root canal procedure and who later brought an

---

[218] Carr v. Strode, 79 Hawaii 475 (1995).

action based on negligence and lack of informed consent after the extraction resulted in complications. In <u>Bernard</u>, Chief Justice Ronald Moon held that:

> "The question of ... causation in an action based on the doctrine of informed consent is to be judged by an objective standard, that is, whether a reasonable person in the plaintiff-patient's position would have consented to the treatment that led to his or her injuries had the plaintiff-patient been properly informed of the risk of the injury that befell him or her."[219]

The current law of informed consent can be stated as follows: Depending on the jurisdiction, it is governed by the tort of battery or negligence, and either the objective patient-oriented standard or the physician-oriented standard will be used to determine disclosure of risks and causation.

---

[219] *Bernard v. Char*, 903 P.2d 667 (Haw. 1995).

SUMMARY POINTS

INFORMED CONSENT

- Informed consent provides the legal authority to treat. Failure to obtain informed consent creates liability either in battery or in negligence.
- Do not confuse a consent form with informed consent.
- The three components of treatment that require disclosure are: procedure, alternatives and risks.
- Determination of what constitutes a material risk that warrants disclosure is a big challenge; serious risks, however remote, should be disclosed.
- Increasingly, the standard of disclosure is the patient-oriented (i.e., what a reasonable patient would want to know under the circumstances) rather than physician-oriented standard.

11

Patients Who Drive: A New Worry for Doctors[220]

What is the physician's legal responsibility when it comes to medical conditions that may pose a driving risk for patients and third parties? This issue has not received the attention it deserves, considering the number of traffic accidents, some of which undoubtedly are related to the medical condition of the driver. Of the few reported court opinions, the trend is to generally hold the doctor liable if there has been a failure to warn the patient-driver of the driving risks. This liability may extend beyond the patient to foreseeable innocent victims of an accident.

MALPRACTICE LAWSUITS

A recent case, <u>McKenzie v. Hawaii Permanente Medical Group, Inc.</u>, puts this issue in the spotlight.[221] The case involved a car suddenly veering across five lanes of traffic, striking Katie, an 11-year-old girl, and crushing her against a cement planter. Katie received multiple traumatic injuries from this accident including physical, cognitive, and emotional impairments. The driver alleged that a prescription medication, Prazosin, caused him to lose control of the car, and that the physician was negligent in two respects: first, in *prescribing* an inappropriate type and dose of medication; and second, in *failing to warn* of the potential side effects that could affect driving ability.

---

[220] Dr. Reina M. Ahern, ObGyn resident, assisted me in the research for this chapter during a senior medical student elective. The original article was published in the June 2003 issue of the Hawaii Medical Journal (Ahern RM and Tan SY. Duty to Third Parties: A New Worry for Doctors. *Hawaii Med J* 2003; 62:123–5). Permission kindly provided by HMJ.
[221] <u>McKenzie v. Hawaii Permanente Medical Group, Inc.</u>, 47 p. 3d 1209 (Haw. 2002).

In analyzing this case, the Hawaii Supreme Court acknowledged that increasing a physician's liability for negligent prescribing might dissuade physicians from prescribing medically necessary medications. The Court emphasized that the risk of tort liability to the individual physician already discourages negligent prescribing; therefore, a physician does not have a duty to third parties where the alleged negligence involves prescribing decisions, i.e., whether to prescribe medication at all, which medication to prescribe, and what dosage to use. On the other hand, physicians have a duty to their patients to warn of potential adverse effects and this responsibility should therefore extend to third parties. Thus, liability would attach to injuries of innocent third parties as a result of failing to warn of a medication's effects on driving — unless a reasonable person could be expected to be aware of this risk without the physician's warning.

Several jurisdictions have previously debated the extent of a physician's duty to third parties under various circumstances. For example, the New Mexico Supreme Court ruled that a physician owes a duty to third parties injured by patients who had been injected with medications, such as narcotics, which are known to affect judgment and driving ability.[222] However, unlike the Hawaii Supreme Court, it later declined to extend a physician's duty to non-patients for prescription-involved situations, after considering the potential impact on malpractice litigation if a physician's duty were to so extend to the general public.[223]

In *Harland v. State of California*, a Court of Appeal affirmed a $3 million verdict for injuries sustained in a head-on collision on the Benicia-Martinez Bridge. The court found that dangerous conditions on the highway bridge was a proximate cause of the collision. The state had failed to properly maintain the bridge for safe use. The driver was a resident at a state-run veterans home. However, the court did not base liability on the state's failure to prevent the veteran from driving his automobile, despite having knowledge a physician at the facility had warned him against driving when he was receiving an antidepressant medication. The court concluded that the plaintiff was not under guardianship, and the state could not simply command him not to drive.[224]

What is the physician's role in warning patients with medical conditions that may lead to hazardous driving? The Iowa Supreme Court has held that a

---

[222] *Wilschinsky v. Medina*, 108 N.M. 511 (1989).
[223] *Lester v. Hall*, 126 N.M. 405 (1998).
[224] *Harland v. State of California*, 75 Cal.App.3d 475 (1977).

physician must warn a patient with a newly diagnosed seizure disorder about the risks of driving. In this case, a patient with a history of a single seizure injured a woman when he suffered a second seizure while driving.[225] As in McKenzie, the third-party plaintiffs filed suit, alleging that the physician negligently failed to warn the patient of potential driving hazards related to his seizure condition. In a similar case, a court in California held that the plaintiff, a third-party accident victim, had a cause of action against two doctors for negligently failing to warn their patient against driving in an uncontrolled diabetic condition that was complicated by a missed abortion.[226]

WARNING OF ADVERSE MEDICATION EFFECTS

An array of documents and resources are available to guide physicians,[227] and the National Transportation Safety Board (NTSB) has recently examined the role of medications in transportation accidents. NTSB proposed to create a list of approved medications for commercial vehicle drivers.[228] Currently, the Food and Drug Administration (FDA) issues driving precautions for numerous medications causing central nervous system depression or hypotension, as well as for other medications that commonly impair driving ability. The list includes the following categories:

- Centrally-acting drugs, including sedatives, antipsychotics and antidepressants
- Antiarrhythmics or other arrhythmogenic agents
- Antihypertensives
- Hypoglycemic agents, such as insulin and sulfonylureas
- Analgesics, especially narcotics
- Topical ophthalmic agents, especially those for treating glaucoma
- Antihistamines and cough medicines
- Other drugs known to cause drowsiness or dizziness

---

[225] Freese v. Lemmon, 210 N.W.2d 576 (Iowa 1973).
[226] Myers v. Quesenberry, 193 Cal.Rptr. 733 (1983).
[227] Drowsy Driving and Automobile Crashes: NCSDR/NHTSA Expert Panel on Driver Fatigue and Sleepiness. National Highway Transportation Safety Administration, December 7, 2000; Seppala, T et al. Drugs, Alcohol, and Driving. Drugs 1979; 17:389–408.
[228] National Transportation Safety Board: Safety Recommendation. January 13, 2000.

Physicians must recognize and explain to their patients how these medications may affect driving ability. For example, driving risks increase with higher medication doses and for people taking more than one drug simultaneously. This risk is greatest soon after initiating the drug regimen. Moreover, these medications may enhance alcohol's deleterious effects. After starting therapy or changing doses, patients should avoid driving until they are aware of the effects of the medication. As always, physicians must exercise their judgment when evaluating these medications, as patients will often react differently. Regardless, warning about the effects of these medications on driving ability is paramount, as is documenting the explanation.

PHYSICIANS' ROLE IN WARNING OF MEDICAL CONDITIONS

The American Medical Association (AMA) has developed a guide to assist healthcare providers who conduct medical examinations of drivers on behalf of licensing authorities.[229] While physicians must use their judgment in determining whether a medical condition affects an individual patient's ability to drive, the AMA has identified specific medical conditions which may impair driving, including:

- Decreased visual acuity, greater than 20/40
- Seizure disorder
- Unstable diabetes mellitus, including those with severe insulin-dependence or persisting ketosis
- Sleep disorders, including narcolepsy and sleep apnea
- Cardiac conditions, including rhythm disturbances, unstable angina, or post-myocardial infarction
- Dementia.

While restrictions may be less onerous for non-commercial vehicle drivers, they are more regulated for commercial vehicle drivers, such as those transporting hazardous materials or busloads of people. The Federal Motor Carrier Safety Administration (FMCSA) has provided specific medical guidelines, including strictly disqualifying commercial vehicle drivers with insulin-dependent diabetes mellitus or those with a blood pressure greater than

---

[229] Doege, TC and Engleberg, AL (eds). *Medical Conditions Affecting Drivers*. American Medical Association. Chicago, IL. 1986.

181/105.[230] States such as Utah have used these guidelines to formulate their own recommendations.[231]

REPORTING TO LICENSING AUTHORITIES

Many motor vehicle agencies have established a medical advisory board (MAB) to evaluate medical issues related to driving and licensure. The members of the MAB, usually a group of appointed physicians, advise the driving examiner with respect to medical criteria and vision standards.[232] If necessary, the examiner may require private physicians to complete a medical questionnaire regarding the patient's driving ability. The questionnaire helps the MAB determine whether the patient is fit to drive. However, it is the driving examiner who makes the ultimate decision regarding licensing qualification.

What if, in the interim period of licensure, a patient's health condition deteriorates to a degree that hinders safe driving? When are doctors legally or ethically obligated to report their patient's condition? In many states, physician reporting of medically impaired drivers is encouraged, and some states actually mandate reporting conditions such as epilepsy.[233] Doctors in California are required to report patients with dementia. Texas, Utah, Florida and Arizona have implemented provisions to immunize healthcare professionals from liability for making judgments regarding patients' ability to drive safely.[234] Other states may make it difficult to report, and doctors may be accused of violating confidentiality should they do so. Doctors should therefore look to their own state laws on this issue before acting.

---

[230] Federal Motor Carrier Safety Administration-Motor Carrier and Driver Laws. Medical Advisory Criteria for Evaluation. 49 CFR Part 391.41.
[231] State of Utah Functional Ability in Driving: Guidelines and Standards for Health Care Professionals, 1992 ed., National Highway Traffic Safety Administration.
[232] Hawaii Revised Statutes §286-4.1: Medical Advisory Board 2001.
[233] Krauss, GL, et al. Individual State Driving Restrictions for People with Epilepsy in the U.S. *Neurology* 2001; 57:1780–5.
[234] Texas Revised Statutes, Health and Safety Code, Title 2, Subchapter H, §12.096 Physician Report; §12.098 Liability 2001; Arizona Revised Statutes, Title 28, Chapter 8, Article 1, §28-3005. Medical or Psychological Reports; Immunity; Definitions 2001; Florida Statutes, Title XXIII, Chapter 322, §322.126 Report of Disability to Department; Content; Use 2002; Utah Code Annotated Title 53 Chapter 3-303. Driver License Medical Advisory Board — Medical Waivers. April 18, 2001.

The AMA's Council on Ethical and Judicial Affairs has a published opinion on reporting patients with driving impairments. First, the physical or mental impairment must "*clearly relate to the ability to drive,*" and "*the driver must pose a clear risk to public safety.*" Appropriate documentation is prudent and necessary, with physicians discussing the risks of driving with the patient and family. Then, depending on the patient's condition, physicians may recommend further treatment options or encourage the patient and family to decide on a restricted driving schedule. Efforts to inform patients, advise them of their options, and negotiate a plan may preempt the need to report. The AMA supports reporting in good faith under strict circumstances:[235]

> "*Physicians should use their best judgment when determining when to report impairments that could limit a patient's ability to drive safely. In situations where clear evidence of substantial driving impairment implies a strong threat to patient and public safety, and where the physician's advice to discontinue driving privileges is ignored, it is desirable and ethical to notify the Department of Motor Vehicles.*"

---

[235] American Medical Association Council on Ethical and Judicial Affairs, Code of Medical Ethics: Current Opinions with Annotations. Impaired Drivers and Their Physicians, Chapter 2.24, 2002–2003 edition.

SUMMARY POINTS

PATIENTS WHO DRIVE: A NEW WORRY FOR DOCTORS

- Physicians may incur liability to patients and non-patient third parties if they fail to warn their patients regarding a medical condition's adverse effect on driving.

- Physicians must be cognizant of a medication's impact on driving ability, and inform their patients of these adverse effects.

- Physicians must consider asking patients to stop driving if the risks posed by a medical condition or medication are substantial.

- Physicians should look to state law to determine whether to report medically impaired drivers.

- It is ethical to report a patient's absolute refusal to stop driving after efforts aimed at convincing the patient of the risk of driving have failed.

12

End-of-Life Issues

It is widely believed that doctors do a poor job at taking care of dying patients. In 1995, the SUPPORT investigators published a multi-center study of over 9,000 critically ill patients and documented that doctors did not regularly address patient wishes regarding do-not-resuscitate (DNR) orders and other end-of-life issues.[236] About 50% of the patients suffered moderate or severe pain. More importantly, treatment approach did not improve despite the doctors being provided prognostic information and patient preferences. Widespread media publicity greeted these results, and prompted a push for corrective measures directed at improving end-of-life care to include the teaching of ethics and pain control to all physicians and trainees.

Many legal myths surround end-of-life issues. Widespread physician fear of liability, often based on erroneous notions of the law, interferes with care of the terminally-ill. Families and their advocates have filed lawsuits against practitioners and hospitals in a number of well publicized cases, but these are rare and they frequently revolve around issues of patient rights rather than malpractice. The areas of greatest potential legal exposure in end-of-life care are:

1. Forgoing life-sustaining treatment
2. Stopping futile treatment
3. Failure to adequately treat pain
4. Assisted suicide and euthanasia.

---

[236] A Controlled Trial to Improve Care for Seriously Ill Hospitalized Patients: The Study to Understand Prognoses and Preferences for Outcomes and Risks of Treatments (SUPPORT). *JAMA* 1995; 274:1541–8.

FORGOING (WITHHOLDING & WITHDRAWING) LIFE-SUSTAINING TREATMENT

Mentally competent adults have the right to accept or reject treatment, even if such decisions result in harm or death. All patients are presumed to have medical decision-making capacity unless determined otherwise. Driven by patient autonomy and dignity, such an approach builds on the doctor-patient relationship of mutual respect, trust and confidentiality. All medical interventions, after all, require implied or explicit informed consent.

Surrogate Decision-Making: Where the patient lacks decision-making capacity, the 'surrogate' as defined by state law, usually a next-of-kin, makes the decision and gives or withholds consent for treatment. In the event that the patient has executed an advance directive (living will), the provisions of that document will govern the type and extent of end-of-life treatment. Advance directives also frequently name the person to whom authority is delegated in the event of incapacity. Such a person is usually called an agent under a durable power of attorney for healthcare decisions. Not infrequently, family members are divided over treatment decisions, and the role of the attending physician is to uncover guilt, ignorance, misconceptions, or denial, and also to provide objective diagnostic and prognostic information. Decisions to forgo treatment require open communication, and an understanding of goals. There is no ethical or legal distinction between withholding and withdrawing treatment, and the doctor, in guiding the discussion, must put the patient's known wishes foremost. If patient preferences are unknown, the next step is to use substituted judgment (what the patient would want if able to communicate), or the best interest test (what a 'reasonable' person would want).

Some states have enacted laws that spell out the procedure for determining the decision-maker when no one has been named. In Hawaii, if there is no designated surrogate, the healthcare team is supposed to assemble all 'interested persons' to determine the surrogate decision-maker, subject to judicial intervention if a consensus is not achieved.

The _Quinlan_ Case: The first and most important legal challenge over the right to discontinue life-sustaining medical treatment is the 1976 _Quinlan_ case.[237] Karen Quinlan was a 21-year-old woman who developed apnea for unclear

---

[237] _In the matter of Karen Quinlan_, 355 A.2d 647 (N.J. 1976).

reasons, suffered irreversible neurological injury, and lapsed into a persistent vegetative state. She became chronically dependent on artificial ventilation in the intensive care unit. Her parents felt that their daughter would not have wanted to live in this way, and opposed the healthcare team's continued use of mechanical ventilation. The case was appealed to the New Jersey Supreme Court, which ruled in favor of the parents:

> "The sad truth however is that she is grossly incompetent and we cannot discern her supposed choice based on the testimony of her previous conversations with friends, where such testimony is without sufficient probative weight. Nevertheless we have concluded that Karen's right of privacy may be asserted on her behalf by her guardian under the peculiar circumstances here present ... If a putative decision by Karen to permit this non-cognitive, vegetative existence to terminate by natural forces is regarded as a valuable incident of her right of privacy, as we believe it to be, then it should not be discarded solely on the basis that her condition prevents her conscious exercise of the choice ... It is for this reason that we determine that Karen's right of privacy may be asserted in her behalf, in this respect, by her guardian and family."

Following the court-ordered removal of the mechanical ventilator, Karen Quinlan was moved to a nursing home, where she remained alive, albeit in a persistent vegetative state, for several more years.

<u>Quinlan</u> stands for the proposition that competent adults have the right to refuse life-sustaining medical treatment, and that the decision to forgo such therapy may be exercised by the patient's family acting as surrogate. The New Jersey Court offered the suggestion that such life-and-death decisions should be resolved by a hospital ethics committee rather than in a court of law. Its analysis took into account the state's legitimate interest in protecting and preserving life but reasoned that under some circumstances, this interest is subservient to patient autonomy.

> "We think that the State's interest contra weakens and the individual's right to privacy grows as the degree of bodily invasion increases and the prognosis dims. Ultimately there comes a point at which the individual's rights overcome the State interest."

Artificial Nutrition & Hydration (ANH): Feeding is a social act of compassion and nurture, and therefore many consider all forms of feeding to be ordinary treatment that should never be withheld. But ANH is quite different from mouth-feeding, and it is more useful to consider whether a proposed treatment is proportionate, i.e., whether the benefits outweigh the burdens, than to attempt to distinguish between ordinary and extraordinary measures. In some instances, forgoing ANH may be the most appropriate and humane course of action.

For a while, a line was drawn to separate ANH from other forms of medical treatment. As the law evolved, this distinction was abolished, and ANH is now regarded as simply another form of medical treatment. In 1990, the U.S. Supreme Court decided *Cruzan v. Director, Missouri Department of Health*,[238] its first 'right-to-die' case. Nancy Cruzan sustained irreversible neurologic damage after a car accident. The Court agreed that states have the power to require a clear and convincing standard before discontinuing life support including ANH:

"Artificial feeding cannot readily be distinguished from other forms of medical treatment ... Whether or not the techniques used to pass food and water into the patient's alimentary tract are termed 'medical treatment,' it is clear they all involve some degree of intrusion and restraint. Feeding a patient by means of a nasogastric tube requires a physician to pass a long flexible tube through the patient's nose, throat and esophagus and into the stomach. Because of the discomfort such a tube causes, 'many patients need to be restrained forcibly and their hands put into large mittens to prevent them from removing the tube.' A gastrostomy tube ... or jejunostomy tube must be surgically implanted into the stomach or small intestine ... Requiring a competent adult to endure such procedures against her will burdens the patient's liberty, dignity, and freedom to determine the course of her own treatment. Accordingly, the liberty guaranteed by the Due Process Clause must protect, if it protects anything, an individual's deeply personal decision to reject medical treatment, including the artificial delivery of food and water." (opinion by Justice O'Connor).

---

[238] *Cruzan v. Director, Missouri Department of Health*, 110 S. Ct. 2841 (1990).

Two recent cases, the _Wendland_ case in California and the _Schiavo_ case in Florida provide additional judicial insights into the issue of when and whether artificial nutrition and hydration can be discontinued.

The _Wendland_ Case:[239] This was a case in which the California Supreme Court articulated the circumstances under which artificial nutrition and hydration may not be withheld. In 1993, 42-year-old Robert Wendland was left comatose after he lost control of his truck. He regained consciousness 14 months later, but was left with hemiparesis. He could not feed by mouth, was incontinent, and could not dress, bathe, or communicate consistently. At best, he was able to draw simple shapes and to follow two-step commands with prompting and cueing. He was able to correctly answer yes or no, and when asked whether he wanted to die, he said no. Totally dependent on tube feeding, he was expected to survive many more years, but without neurologic recovery.

In 1995, his wife Rose refused to authorize reinsertion of the dislodged feeding tube. Robert did not execute an advance directive, but had apparently made statements to the effect that he would not want to live in a vegetative state. The wife's decision was supported by his daughter and brother, the hospital's ethics committee, county ombudsman, and court appointed counsel. But the patient's mother, Florence, disagreed and went to court to request reinsertion of the feeding tube.

In an unanimous 6-0 decision, the California Supreme Court ruled that Robert Wendland's tube feedings could not be discontinued, although by the time the Court handed down its ruling, Robert had died from pneumonia. The Court determined that Wendland's statements were not clear and convincing because they did not address his current condition, were not sufficiently specific, and were not necessarily intended to direct his medical care. Ruling that his wife Rose Wendland had failed to provide sufficient evidence that her decision was in her husband's best interests, the Court declared that its ruling applied to patients who were "minimally conscious" and not to those who were terminally ill.

Brief Comments Regarding _Wendland_: This was a case where the patient was neither terminal nor in a persistent vegetative state. He exhibited residual,

---

[239] _Wendland v. Wendland_, 28 P.3d 151 (Cal. 2001).

even if minimal, preservation of cognitive function. In a conscious patient like Robert Wendland, a conservator may withhold or withdraw life-sustaining treatment only if there is clear and convincing evidence that the patient wished to refuse life-sustaining treatment or that to withhold such treatment would have been in his best interest. According to the California Supreme Court, these conditions were not met.[240]

The *Terri Schiavo* Case:[241] In 1990, Terri Schiavo, then a young woman of 26, lapsed into a non-cognitive state after suffering a catastrophic cardiac event that was possibly related to hypokalemia. The underlying cause was never made clear, but there were allegations that she received negligent medical care for an eating disorder. Her husband Michael Schiavo successfully filed a medical malpractice lawsuit and was awarded damages of $1.5 million. Since the incident, she had to be fed through a feeding tube. Terri Schiavo left no advance directive, but her husband said she would rather die than be kept alive artificially, and he therefore wanted to stop her tube feeding. However, her parents Bob and Mary Schindler, wished to keep her alive, even volunteering to be fully responsible for her care.

After a long legal battle, the husband succeeded in persuading the appellate court to his viewpoint. The Court found clear and convincing evidence that Terri Schiavo would not have wanted to live in that manner, and both the Florida Supreme Court and U.S. Supreme Court refused to review the case. Accordingly, on October 15, 2003, her tube feeding was stopped. Her parents appealed to Governor Jeb Bush to intervene, calling into question the experts' diagnosis that she was in a persistent vegetative state (PVS). They insisted that her eyes were opened, and she appeared to respond with smiles to the sound of her mother's voice. They were also able to get fifteen doctors to testify that she could and would improve.

On October 21, 2003, the Florida legislature passed an emergency bill, popularly known as 'Terri's Law,' that allowed the governor to order reinsertion of the feeding tube. The Florida Supreme Court, however, ruled Terri's Law to be unconstitutional, and the U.S. Supreme Court again refused to hear the appeal. On March 18, 2005, Terri Schiavo's feeding tube was again removed.

---

[240] Lo, B *et al.*, The *Wendland* Case — Withdrawing Life Support from Incompetent Patients Who Are Not Terminally Ill. *New Engl J Med* 2002; 346:1489–93.
[241] Annas, GJ. "Culture of Life" Politics at the Bedside — The Case of Terri Schiavo. *New Engl J Med* 2005; 352:1710–5.

The U.S. Congress met in emergency session on Sunday, March 20, 2005 and passed a bill entitled: "*For the relief of the parents of Theresa Marie Schiavo.*" The new law instructed the Federal District Court to "*determine de novo any claim of a violation of any right of Theresa Marie Schiavo ... notwithstanding any prior State court determination ...*" However, the U.S. District Court in Florida (Judge James D. Whittemore) denied an emergency request by the parents to reinsert the feeding tube, because they had failed to demonstrate "*a substantial likelihood of success on the merits*" — the standard needed to issue a temporary restraining order.

Teri Schiavo died at 10 a.m. on March 31, 2005, 13 days after withdrawal of the tube feeding. She was 41-years-old. The autopsy revealed severe brain atrophy; she was indeed in a permanent vegetative state with no chance of recovery.

Brief Comments Regarding *Schiavo*: This heart-wrenching case that incredibly reached all three branches of the U.S. government is without precedent. The case polarized the nation, with as many calling for re-insertion of the feeding tube as those siding with the husband. The case was emotionally charged, in part because Terri's true wishes as represented by her husband, Michael, were questioned by many, though not by the courts. The case is about 'who' (has decision-making capacity), 'what' (can be removed, in this case artificial nutrition and hydration) and 'when' (coma versus PVS versus terminally-ill). It is a reminder of the importance of executing an advance directive and appointing a surrogate decision-maker. It is not about medical malpractice, although ironically, a malpractice suit was filed and won after her initial medical mishap, with damages of $1.5 million awarded to purportedly cover a life expectancy of 50 years.

The law regarding decisions to forgo life-sustaining treatment can be summarized as follows:

(1) An adult with decision-making capacity may choose to forgo such therapy.
(2) An advance directive serves to legally express the patient's wishes when he or she is no longer able to do so.
(3) In the event there is no advance directive, the patient's surrogate decision-maker may make the decision.
(4) Artificial nutrition and hydration is considered a form of medical treatment that can be withdrawn or withheld under the appropriate circumstances.

(5) Disputes tend to arise when the legitimate decision-maker is not well identified; the wishes of the patient are unclear; the patient is not terminally ill; and the decision concerns 'ordinary treatment' such as artificial nutrition and hydration.

STOPPING FUTILE TREATMENT

Medical futility is a relatively new concept in medical ethics. A Medline search of the 1966–91 literature yielded only one citation, but in the three-year span from 1997 to 1999, there were 217. There are four general definitions of medical futility: (1) Intervention has no pathophysiologic rationale; (2) Intervention has already failed in the patient; (3) Maximal treatment is failing; and (4) Intervention will not achieve the goal of care. Schneiderman et al. offer the best working quantitative principle of futility: *"When physicians conclude (either through personal experience, experiences shared with colleagues or consideration of reported empiric data) that in the last 100 cases, a medical treatment has been useless, they should regard that treatment as futile."* [242]

Whereas <u>Quinlan</u> and <u>Cruzan</u> were about discontinuing life-sustaining treatment, futility issues are the opposite, being about family insistence on treatment when it can no longer be beneficial. Although judgments about futility may be value-laden, proponents of the concept of medical futility argue that all medical decisions are value-laden, and defining futility is simply an honest attempt at establishing reasonable and objective guidelines. Recognizing that the issue is controversial when the various parties have discrepant values and treatment goals, the AMA has proposed a process-based approach that included options such as transfer to another physician or another institution.[243] When all options have been exhausted, the AMA believes that it is ethical to unilaterally cease futile intervention.

The most common hospital scenario where issues of medical futility are raised is in the intensive care unit, where a patient may be on life-support for an extended time period, and the healthcare team believes recovery is highly unlikely. Another example is cardiopulmonary resuscitation (CPR) in patients with irreversible conditions such as dementia, end-stage liver disease

---

[242] Schneiderman, L et al. *Ann Int Med* 1990; 112:949–54.
[243] Report of the Council on Ethical and Judicial Affairs: Medical Futility in End-of-Life Care. *JAMA* 1999; 281:937–41.

or a terminal malignancy. In these cases, patients or family members may be unrealistic and demand that everything be done, although the life-support measures can no longer confer any meaningful benefit. Consultations with the hospital ethics committee may offer some help in these situations, if only because an outside objective voice(s) may help to educate and convince family members of the futility of continued treatment.

Futility issues typically arise in cases of irreversible coma or persistent vegetative states. Such patients are fully alive under the law, and must be distinguished from brain dead patients, where there is no further obligation to treat. Brain death is defined by statute, and California's language is typical: "*Irreversible cessation of all functions of the entire brain, including the brain stem.*"[244]

The Law: What does the law have to say about futile treatment? Statutory law is at best relatively indirect. Unilateral decisions by the healthcare provider to discontinue or withhold futile life-sustaining treatment without the consent of the patient or family is apparently at odds with New York's law on cardiopulmonary resuscitation (CPR).[245] Section 2962, paragraph 1 states: "*Every person admitted to a hospital shall be presumed to consent to the administration of cardiopulmonary resuscitation in the event of cardiac or respiratory arrest, unless there is consent to the issuance of an order not to resuscitate as provided in this article.*" However, Hawaii's law appears to support physician decisions based on medical indications: "*This chapter shall not authorize or require a health-care provider or institution to provide healthcare contrary to generally accepted health-care standards applicable to the health-care provider or institution.*"[246]

Medical futility cases have reached the courts, with three achieving notoriety. The first involved Helga Wanglie, an elderly woman in a persistent vegetative state in whom the healthcare providers attempted to discontinue artificial ventilation against the wishes of her husband. The Minnesota court ruled that Mr. Wanglie was competent to make decisions regarding the healthcare of his wife, i.e., to continue life-sustaining artificial ventilation against

---

[244] California Health and Safety Code §7180.
[245] N.Y. Pub. Health Law §§2960–2978.
[246] Hawaii Revised Statutes §327E-13(d).

the doctors' advice, and managed to sidestep the issue of medical futility altogether.[247]

The next case was that of <u>Baby K</u>.[248] Physicians petitioned to withdraw ventilator support that had been provided for an anencephalic baby, arguing that: *"A requirement to provide respiratory assistance would exceed the prevailing standard of medical care ... because any treatment of their condition is futile, the prevailing standard of medical care for infants with anencephaly is to provide warmth, nutrition and hydration."* The mother insisted that care must be continued for her infant. The Virginia court also chose to deflect the futility issue, and instead used EMTALA (Emergency Medical Treatment and Active Labor Act — popularly known as the anti-dumping law) — to require all emergency rooms to provide *"treatment necessary to prevent the material deterioration of the individual's condition."* The case was affirmed by the Fourth Circuit Court of Appeals, which held that the hospital was not authorized to decline to provide stabilizing treatment which it considered morally and ethically inappropriate.[249]

The third case in this triad, <u>Gilgunn v. Massachusetts General Hospital</u>, was won by the doctors.[250] It is the only lawsuit that challenged a doctor's judgment <u>after</u> futile treatment was withheld. Catherine Gilgunn, a 71-year-old woman with multiple medical problems was hospitalized for uncontrolled diabetes and a hip fracture. During her hospital course, she sustained two grand mal seizures that could not be controlled with anti-convulsants, and irreversible coma ensued.

Joan, her 30-year-old daughter, was the surrogate decision-maker, and she made it clear to the physicians that her mother had always wanted every medical intervention, even those that are unlikely to be effective. When asked to consult on the advisability of CPR, Father Edwin Cassem, chair of the hospital ethics committee, believed that the daughter's opinion was irrelevant because CPR was not a genuine therapeutic option. The attending physician entered a DNR order, but upon strong protest from Joan Gilgunn, revoked the order

---

[247] <u>In re Helga Wanglie</u>. Fourth Judicial District (Dist. Ct., Probate Ct. Div.), PX-91-283. Minnesota, Hennepin County.
[248] <u>In re Baby K</u>, 832 F. Supp. 1022 (4th Cir., Va. 1993).
[249] <u>In re Baby K</u>, 16 F.3d 590 (4th Cir., Va. 1994).
[250] <u>Gilgunn v. Massachusetts General Hospital</u>, Super. Ct. Civ. Action No. 92-4820, Suffolk Co., Mass., verdict, 21 April, 1995.

two days later. The following month, a new attending physician asked for another ethics review. Father Cassem again endorsed a DNR order, because CPR would be *"medically contraindicated, inhumane, and unethical."* The attending physician then began to wean Mrs. Gilgunn from the respirator, and three days later, she died without CPR.

Joan Gilgunn filed suit against the hospital. The defendant's expert, John Paris, S.J., relied upon the position paper of the American Thoracic Society (ATS) on life-sustaining treatment. According to the ATS, life support *"can be limited without the consent of patient or surrogate when the intervention is judged to be futile . . . that would be highly unlikely to result in a meaningful survival of the patient . . . Survival in a state of permanent loss of consciousness . . . may be generally regarded as having no value for such a patient."*

After two hours of deliberation, the jury found that if competent, the patient would have wanted CPR and would have wanted ventilation until death. But it further found such treatment would be futile. In its verdict issued on April 21, 1995, the jury found no negligence.

The tide appears to be turning. The law may finally be cognizant of the concept of medical futility, and be more willing to defer to the clinical judgment of healthcare professionals regarding the utility of life-sustaining treatments.

FAILURE TO ADEQUATELY TREAT PAIN

Pain management and comfort care become primary treatment goals once cure is impossible. Many dying patients fear that pain will not be aggressively treated, and studies repeatedly show that physicians do not adequately treat pain in terminally-ill patients. In addition to providing pain relief, physicians should communicate their plans regarding palliative care by using open-ended questions, screening for unaddressed spiritual concerns, and listening actively and empathically.[251] Both the legal and ethical literature consistently support the prescription of narcotics in sufficient dosage to effectively relieve pain, even if death is hastened because of respiratory depression or other systemic effect. This phenomenon is termed the 'double effect.'

---

[251] Lo, B et al. Discussing Palliative Care With Patients. *Ann Int Med* 1999; 130:744–9.

In a recent California trial that received widespread media coverage, an Alameda County jury turned in a verdict against an internist charged with elder abuse and reckless negligence because he failed to give enough pain medication to a Hayward man dying of cancer.[252] Under California law, death of a plaintiff extinguishes a claim for pain and suffering. The case was therefore brought under the elder-abuse law, although the burden of proof was higher, requiring a reckless rather than a simple negligence standard.

The case involved William Bergman, an 85-year-old retired railroad worker with lung cancer, who was admitted to Eden Medical Center in early 1998. The lawsuit alleged that the treating physician was reckless in not prescribing effective medication for Bergman who complained of severe back pain. Bergman stayed at the hospital for six days as nurses consistently charted his pain in the 7–10 range. On the day of discharge, his pain was at level 10. He died at home shortly thereafter.

After four days of deliberation, the jury, in a 9–3 vote, entered a guilty verdict, and awarded $1.5 million in general damages. This amount was subsequently reduced to $250,000 because of California's cap on non-economic damages. Eight jurors wanted to award punitive damages, as they believed that the doctor had acted with malice or had intentionally caused emotional distress. However, no punitive damages were assessed because nine votes were needed. The hospital had settled privately with the family before trial. The guilty verdict came despite defense expert testimony that the treatment provided was reasonable and would be the same as that provided by 95% of all internists.

Bergman's family had earlier filed a complaint with the California Medical Board, which took no action because it did not find clear and convincing evidence of a violation of the Medical Practice Act, although the board's medical consultant concluded that the hospital's pain management was inadequate. The verdict prompted a state legislator to author a bill that would require all California doctors to take a pain management course.

The <u>Bergman</u> case is notable for being the first of its kind, and squarely puts physicians on notice regarding their duty to adequately provide pain relief. The closest previous case of liability for failure to treat pain involved a

---

[252] *Bergman v. Eden Medical Center*, No. H205732-1 (Sup. Ct. Alameda Co., Cal., 2001).

nursing home's failure to administer pain medications that had been ordered by the doctor.[253]

ASSISTED SUICIDE & EUTHANASIA

In 1988, a short article appeared in *JAMA* entitled "It's Over, Debbie." The article was purportedly written by a physician who deliberately injected an overdose of morphine to end the suffering of a young woman dying of ovarian cancer.[254] The case ignited an uproar, and paved the way for an intense national debate on euthanasia.

The debate was fueled in the early 1990s with reports that Dr. Jack Kervokian, a Michigan pathologist, had assisted over a hundred patients in committing suicide. Initially escaping conviction, he was finally found guilty of second-degree murder in 1999, and imprisoned for 10–25 years after he gave a lethal injection to a middle aged man suffering from Lou Gehrig's disease. Although attempted suicide itself is no longer a criminal offence, assisting someone to commit suicide remains a felony. For example, Hawaii law considers that: *"A person commits the offense of manslaughter if ... he intentionally causes another person to commit suicide."*[255] The prohibition is even broader in Kansas.[256] There, it is a criminal act if there is the intent and purpose of assisting another person to commit or to attempt to commit suicide, knowingly either by providing the physical means, or participating in a physical act by which another person commits or attempts to commit suicide.

In 1997, the United States Supreme Court took on two cases that challenged these laws as applied to physician-assisted suicide. In New York, as in most States, it is a crime to aid another to commit or attempt suicide, yet patients may refuse lifesaving medical treatment. Thus in Vacco v. Quill,[257] the U.S. Supreme Court was asked whether New York's prohibition on assisting suicide violates the Equal Protection Clause of the 14th Amendment. The

---

[253] *Estate of Henry James v. Hillhaven Corp.*, Super Ct. Div. 89 CVS 64 (Hertford Cty., N. C. November 20, 1990).
[254] Anonymous. It's Over Debbie. *JAMA* 1988; 259:272.
[255] Hawaii Revised Statutes §707-702.
[256] *Kan. Stat. Ann.* §21-3406.
[257] *Vacco v. Quill*, 117 S. Ct. 2293 (1997).

Court held that it did not:

"Neither New York's ban on assisting suicide nor its statutes permitting patients to refuse medical treatment treat anyone differently than anyone else or draw any distinctions between persons. Everyone, regardless of physical condition, is entitled, if competent, to refuse unwanted lifesaving medical treatment; no one is permitted to assist a suicide. Generally speaking, laws that apply evenhandedly to all "unquestionably comply" with the Equal Protection Clause."

The Court reasoned that the distinction between assisting suicide and withdrawing life-sustaining treatment *"comports with fundamental legal principles of causation and intent."* When a patient forgoes life-sustaining treatment, death results from the underlying fatal disease; but if a patient ingests lethal medication prescribed by a physician, he is killed by that medication. Furthermore, in honoring a patient's preference, a physician merely ceases futile or degrading treatment where benefit can no longer accrue. Patients who refuse life-sustaining treatment may not specifically intend to die. Indeed, they may fervently wish to live, but to do so free of unwanted medical technology. When a doctor provides aggressive palliative care, the drugs used, e.g., morphine, may hasten a patient's death, but the physician's purpose and intent is to ease his patient's pain, and not to cause or accelerate his or her death. Contrast this with the doctor who assists a suicide, who, *"however, must, necessarily, and indubitably, intend primarily that the patient be made dead."* The U.S. Supreme Court concluded:

"The law has long used actors' intent or purpose to distinguish between two acts that may have the same results ... Put differently, the law distinguishes actions taken "because of" a given end from actions taken "in spite of" their unintended but foreseen consequences ... When General Eisenhower ordered American soldiers onto the beaches of Normandy, he knew that he was sending many American soldiers to certain death ... His purpose, though, was to ... liberate Europe from the Nazis."

In a companion case, the Supreme Court decided *Washington v. Glucksberg*,[258] a similar right-to-die case where the issue was whether the protections of the Due Process Clause included a right to commit suicide with another's assistance. Washington law provides that: *"A person is guilty of promoting a suicide attempt when he knowingly causes or aids another person to attempt suicide. Promoting a suicide attempt is a felony, punishable by up to five years' imprisonment and up to a $10,000 fine."*[259]

The litigants were three terminally-ill plaintiffs and Compassion in Dying, a non-profit organization. They asserted that the right to physician-assisted suicide is a fundamental liberty interest protected by the Due Process Clause, and their position found support in the Court of Appeals for the Ninth Circuit. However, the United States Supreme Court unanimously reversed, holding that the relevant Washington statute did not violate the Fourteenth Amendment, either on its face or as applied to competent, terminally ill adults who wish to hasten their deaths by obtaining medication prescribed by their doctors.

The Court acknowledged the various legitimate interests of the state that included preserving human life, preventing suicide, protecting the integrity and ethics of the medical profession, and protecting vulnerable groups, including the poor, elderly and disabled persons, from abuse, neglect and mistakes. The state may also fear that permitting assisted suicide will initiate the path to voluntary and perhaps even involuntary euthanasia. The Constitution requires that Washington's assisted-suicide ban be rationally related to legitimate government interests, and the Court concluded that *"This requirement is unquestionably met here..."*

Thus, these two seminal cases put to rest the assertion that there exists a constitutional right to physician-assisted suicide. However, the Court did leave open the possibility that states may continue to debate the issue:

> *"Throughout the Nation, Americans are engaged in an earnest and profound debate about the morality, legality, and practicality of*

---

[258]*Washington v. Glucksberg*, 117 S. Ct. 2258 (1997).
[259]Wash. Rev. Code 9A.36.060(1)(1194).

physician assisted suicide. Our holding permits this debate to continue, as it should in a democratic society."

Oregon's Death with Dignity Act:[260] In 1994, the electorate in Oregon voted in favor of Measure 16, and the state became the first, and still only, to legalize physician assisted suicide (PAS). Supporters of Measure 16 narrowly outnumbered opponents in Oregon after earlier similar initiatives failed in Washington and California. With Measure 16 encoded in Oregon's Death with Dignity Act, physicians in Oregon may now legally prescribe a lethal dose of a barbiturate to a competent terminally-ill resident with the specific intent to end life. There are however, built-in safeguards to prevent abuse.

During Oregon's first year with legalized PAS, 23 patients received prescriptions for lethal medications, and 15 died after taking the lethal medications. Use of the lethal medication was associated with concern about loss of autonomy or control of bodily functions, rather than with fear of intractable pain (only 1 case) or financial loss (no case).[261] During the second year,[262] 27 patients died after ingesting prescribed lethal medications. Again, the reasons given were loss of autonomy, loss of control of bodily functions, an inability to participate in activities that make life enjoyable, and a determination to control the manner of death. A review of Oregon's experience from October 1997 through August 1999 revealed that 5% of physicians received a total of 221 requests for lethal medications. Only 18% of patients received prescriptions, and 10% died from taking the prescribed medication. Twenty percent of the patients were depressed and none of these patients received prescriptions. Substantive palliative interventions led 46% of patients to change their minds about assisted suicide.[263] A recent perspective nicely updates the politics and experiences surrounding this law.[264]

---

[260] Oregon Death with Dignity Act. Ore. Rev. Stat. §§127.800-127.897 (1994).
[261] Chin, AE et al. Legalized Physician-assisted Suicide in Oregon — The First Year's Experience. N Engl J Med 1999; 340:577–83.
[262] Sullivan, AD et al. Legalized Physician-assisted Suicide in Oregon — The Second Year. N Engl J Med 2000; 342:598–604.
[263] Ganzini, L et al. Physicians' Experiences with the Oregon Death with Dignity Act. N Engl J Med 2000; 342:557–63.
[264] Okie, S. Physician-Assisted Suicide — Oregon and Beyond. N Engl J Med 2005; 352:1627–30.

Several states have rebuffed attempts at legalizing PAS, including Michigan, Hawaii and Maine. Whether Oregon will remain the only state with legalized PAS remains to be seen. Polls indicate that a significant percent of the public and healthcare professionals favor some form of legalized PAS. The American Medical Association, however, is on record as being against PAS.

The Bush administration is opposed to Oregon's statute, and the U.S. Supreme Court has agreed to hear arguments on the case, *Oregon v. Gonzales*. The Court's decision is expected in mid-2006.

Criminal Prosecution: In addition to statutorily prohibited assistance with suicide, other medical decisions at life's end may raise the specter of criminal sanctions. The prosecutor may feel that there has been a violation of the criminal code (a homicide) when confronted with a patient whose death was hastened.

Homicide is an act causing the death of another person with criminal intent and without legal justification. Under this definition, the *actus reus* element is satisfied where there is a positive act, or where an omission to act occurs in the presence of a legal duty. Thus, a lifeguard's deliberate failure to save a drowning person is an omission. By failing to act when the lifeguard has a legal duty to, he can be charged for the crime of homicide. A casual beachgoer, on the other hand, owes no such duty to one who is drowning, and can afford to stand by with impunity with no risk of criminal sanction. Here, legal duty is to be distinguished from moral duty. One may feel ethically obliged to save a drowning person, but commits no crime by not doing so.

Like all private citizens, the physician is forbidden by the criminal law from performing any act that leads to the unlawful taking of a human life. It matters not if the patient requested or consented to the killing, or if the killing was motivated by mercy (Oregon's law being the sole exception). Whereas informed consent authorizes the physician to begin medical or surgical therapy, consent is never accepted as a defense to homicide. Such killings undoubtedly occur, but are believed to be rare. Reported cases include the arrest of Dr. John Kraai, a 76-year-old physician, for using insulin in the mercy killing of his long-time friend and patient with Alzheimer's disease.[265] In two other physician-defendant murder cases under similar facts, both defendants

---

[265] A Doctor's Fatal Remedy, *Times* Magazine, September 9, 1985 p. 45.

were acquitted, with the jury finding that the cause of death could not be established with sufficient certainty. One involved the use of intravenous potassium chloride,[266] and the other, alleged injection of air[267] to cause death in a cancer patient.

As a health professional entrusted with saving the lives of patients, the physician's omission of a life-saving act may be a breach of his or her legal duty. Regardless of whether it is classified as an omission or a positive act, withholding or withdrawing treatment that leads to a patient's death raises the specter of criminality. Until the *Barber* case, no physician had been prosecuted for choosing to terminate life-preserving or emergency treatment in a hopelessly ill patient.

In *Barber v. Superior Court*, California decided its first case where criminal charges of murder were brought for the withdrawal of life support.[268] In 1982, two California doctors were charged with murder and conspiracy to commit murder for discontinuing intravenous fluids and nutrition from a comatose patient. The patient was a 55-year-old security guard who suffered a cardiopulmonary arrest following surgery for intestinal obstruction. Irreversible brain damage resulted, leaving him in a vegetative state. His family requested that life support measures and intravenous fluids be discontinued. The doctors complied, and the patient died six days later. After a preliminary pre-trial hearing, the magistrate dismissed the charges, but a trial court reinstated them.

The Court of Appeals, however, dismissed the charges. It viewed the defendant's conduct in discontinuing intravenous fluids as an omission rather than an affirmative action, and found that a physician has no duty to continue treatment once it is proven to be ineffective. The Court dismissed the prosecutor's contention that unlike the respirator, fluids and nutrition represented ordinary care and therefore should never be withheld. In rejecting the distinction between ordinary and extraordinary care, it adopted the benefit-burden or proportionate treatment test, stating: *"we view the use of an intravenous*

---

[266] *People v. Montemarano*, Indictment No. 37707, Nassau County Court, NY (1974).
[267] As reported in President's Commission for the Study of Ethical Problems in Medicine and Biochemical Research: Deciding to Forego Life-Sustaining Treatment (US Government Printing Office, Washington, 1983) at pp. 34–35.
[268] *Barber v. Superior Court*, 147 Cal.App.3d 1006 (1985).

administration of nourishment and fluids, under the circumstances, as being the same as the use of the respirator or other form of life support equipment."

The Court concluded that: *"the petitioners' omission to continue treatment under the circumstances, though intentional and with knowledge that the patient would die, was not an unlawful failure to perform a legal duty."* And since no criminal liability attaches for failure to act unless there was a legal duty to act, it issued a writ of prohibition restraining the Superior Court of Los Angeles County from taking any further action on the matter.

To reach the same result, the Court could have chosen to deal with the issue as one of proximate cause. It could have found that the physicians' conduct was not the agent responsible for the patient's death, which resulted from natural causes set into motion neither by the patient, the family nor the physicians. Absent the causation element, the homicide charge would then become untenable. In ruling for the physician defendants, the Barber court unanimously confirmed and extended the judicial philosophy first enunciated in Quinlan respecting the traditional medical practice of joint decision-making by physicians and families of incompetent patients.

Barber remains the only case of its kind, and it is unlikely that the law will prosecute similar cases. As articulated by the Quinlan Court, Barber's greatest impact may have been *"to free physicians, in the pursuit of their healing vocation, from possible contamination by self-interest or self-protection concerns which would inhibit their independent medical judgments for the well being of their dying patients."*[269]

---

[269] In the matter of Karen Quinlan, 355 A.2d 647, 668 (N.J. 1976).

SUMMARY POINTS

END-OF-LIFE ISSUES

- A patient's right to forgo life-sustaining treatment extends to the withholding or withdrawal of artificial nutrition and hydration.

- Where life-sustaining measures can no longer confer any meaningful benefit and a state of medical futility is reached, physicians are obligated to explain the situation to family members in order to cease further treatment.

- Physicians should be attentive to end-of-life care and especially comfort care. They should not hesitate to treat pain aggressively, even if death is unintentionally hastened ('double effect').

- Except in Oregon, physician-assisted suicide and euthanasia are illegal acts under current law, and physicians who violate the law do so at their own peril.

13

Complementary and Alternative Medicine[270]

Physicians are trained to utilize evidence-based medicine, which determines safe and efficacious patient care through randomized controlled trials. However, many patients seek largely unproven means outside of conventional biomedicine to heal themselves and better their health. Understanding the medico-legal implications of unconventional medicine is therefore an important component of understanding and avoiding malpractice.

Complementary and Alternative Medicine (CAM) has been defined as medical interventions not taught widely in U.S. medical schools or generally available in U.S. hospitals. CAM covers a broad range of healing philosophies, approaches and therapies that are outside mainstream Western medicine. It is estimated that up to 42% of the population uses some form of alternative medicine for a variety of reasons. It is perceived as natural and therefore safe and is enticing because it promotes overall well-being in addition to treating a specific illness.[271]

As the American public's interest in and utilization of CAM increases, state legislatures and professional medical organizations have begun to license practitioners, establish standards of practice and protect healthcare consumers by creating standardized credentials as markers of competence. In

---

[270] Dr. Stefanie Park, Internal Medicine resident, assisted me in the preparation of this chapter during a senior medical student elective.
[271] Eisenberg, DM et al. Unconventional Medicine in the United States. *N Engl J Med* 1993; 328: 246–52; Eisenberg, DM et al. Trends in Alternative Medicine Use in the United States, 1990–1997. Results of a Follow-up National Survey. *JAMA* 1998; 280:1569–75; Astin, JA. Why Patients Use Alternative Medicine: Results of a National Study. *JAMA* 1998; 279:1548–53; Kaptchuk, TJ and Eisenberg, DM. The Persuasive Appeal of Alternative Medicine. *Ann Int Med* 1998; 129:1061–5.

1992, the National Institutes of Health formed an Office of Alternative Medicine in order to evaluate alternative medicine remedies. In 1998, the National Center for Complementary and Alternative Medicine (NCCAM) was instituted as the Federal Government's lead agency for scientific research on complementary and alternative medicine.[272]

CAM comprises many modalities, with over-the-counter 'dietary supplements' being the most popular. Provisions of the Dietary Supplement Health and Education Act of 1994 exempt such supplements from pre-market safety evaluation by the Food and Drug Administration so long as manufacturers make no claims linking the supplements to a specific disease.[273] Consequently, dietary supplements are widely available and may vary tremendously in content, safety and quality.

The scope of practice for CAM providers is defined and limited by state rather than federal statutes, and enforced by regulatory bodies. Such statutes ensure that providers will offer services within their competence, as CAM practitioners still lack the consistency and uniformity of training and credentialing of other healthcare professions.[274] However, there has been significant movement forward: Chiropractors are now licensed in all states, non-MD acupuncturists in 42 states, and massage therapists in 25 states. A minority of states also issue licenses to naturopaths and homeopaths.[275] The state of Hawaii, for example, recognizes chiropractors, acupuncturists, massage therapists and naturopaths.[276]

MALPRACTICE OF CAM PRACTITIONERS

CAM treatments are generally noninvasive, and there are fewer than 50 indemnity insurers in the country for chiropractors, massage therapists and acupuncturists. They underwrite approximately 5% of the total medical malpractice insurance market. Most claims relate to minor injuries or to

---

[272] NCCAM website: www.nccam.nih.gov
[273] Dietary Supplement Health and Education Act of 1994, Public Law 103–417.
[274] Cohen, MH. Legal Issues in Complementary and Integrative Medicine. A Guide for the Clinician. Med Clin N Amer 2002; 86:185–96.
[275] Eisenberg, DM et al. Credentialing Complementary and Alternative Medical Providers. Ann Int Med 2002; 137:965–73.
[276] Chiropractors: Hawaii Revised Statutes §442; Acupuncture: Hawaii Revised Statutes §436; Massage Therapy: Hawaii Revised Statutes §452; Naturopathy: Hawaii Revised Statutes §455.

sexual misconduct.[277] For a successful malpractice suit, the plaintiff must establish breach of duty. The plaintiff must prove that the CAM provider deviated from the standard of care ordinarily exercised by a similarly situated CAM practitioner. Standards of care can vary widely among CAM practitioners, whose treatments are often highly individualized.[278] Therefore, breach of duty would be difficult to prove. Common allegations are failure to refer to a medical doctor, and practicing outside the scope of CAM and venturing into traditional western medical practice. In the latter instance, the CAM practitioner will be held to the higher standard of the medical practitioner.

The most likely CAM practitioner to be sued for malpractice is the chiropractor. In one instance, a plaintiff alleged that he was led to believe that chiropractic manipulation would help his diabetes. The court found in favor of the plaintiff and awarded damages.[279] In another, the plaintiff successfully sued a chiropractor for failing to take X-rays and to refer to a medical doctor. The court held that the defendant fell below the standard of care, as the state licensing board required physician referral when the problem extended beyond the limits of chiropractic practice.[280]

The injured party does not always prevail. In *Miyamoto v. Lazo*, the plaintiff claimed that Dr. Lazo, a chiropractor, negligently treated his injuries from a car accident. The patient was taking Coumadin, and developed a hematoma in his left shoulder following chiropractic treatment. This was complicated by neuropathy in his left hand when the hematoma expanded and required surgical drainage. The jury, however, found Dr. Lazo not liable because of evidence that a hematoma could spontaneously arise in someone on an anticoagulant.[281]

Though uncommon, lawsuits occasionally do get filed against other CAM practitioners. In *Wallman v. Kelley*,[282] the plaintiff who contracted chemical hepatitis claimed negligence and breach of implied warranty claims against a

---

[277] Studdert, DM et al. Medical Malpractice Implications of Alternative Medicine. *JAMA* 1998; 280:1610–5; Angell, M and Kassirer, JP. Alternative Medicine — The Risks of Untested and Unregulated Remedies. *N Engl J Med* 1998; 339:839–41.
[278] Cohen, MH. *Complementary and Alternative Medicine. Legal Boundaries and Regulatory Perspectives*. Baltimore, Johns Hopkins University Press, 1998.
[279] *Wengel v. Herfert*, 473 N.W.2d 741 (Mi. 1991).
[280] *Salazar v. Ehmann*, 505 P.2d 387 (Co. 1972).
[281] *Miyamoto v. Lum and Lazo*, 84 P.3d 509 (Haw. 2004).
[282] *Wallman v. Kelley*, 976 P.2d 330 (Co. 1998).

seller of Chinese herbal medicine. The court held that the defendant was not liable as the plaintiff failed to prove causation or give timely notice of suit.

DOCTORS WHO PRACTICE CAM

By definition, CAM is not taught as tried-and-true therapy in medical schools. Therefore, the very implementation of non-standard therapy may be equated with substandard care, and physicians must be very cautious when practicing outside conventional medicine. One appellate judge has warned that *"Currently, the law does not encourage medical doctors to stray from the pack (because) it is well settled that in medical malpractice actions, the question of negligence must be decided by reference to relevant medical standards of care...."* [283]

Lack of informed consent is usually the basis of lawsuits against physicians who practice CAM. In <u>Charell v. Gonzalez</u>, a cancer patient refused treatment by oncologists and opted instead for nutritional therapies. Her cancer metastasized, leading to blindness and back problems. The patient alleged negligence and failure to warn of risks. The jury found the physician 51% liable for departure from standard of care and lack of informed consent, which resulted in patient injury. The plaintiff was found to be 49% at fault for choosing to ignore the recommendations of her oncologists.[284]

In <u>Moore v. Baker</u>, a patient attempted to sue her neurologist for failure to disclose the option of EDTA chelation therapy as an alternative to surgical treatment for coronary blockages. In this case, the patient had undergone a carotid endarterectomy and during the recovery period, a blood clot developed, impeding blood flow to the brain and causing brain damage. The plaintiff alleged that EDTA chelation therapy was as effective as surgery and was less risky, and that the physician had failed to inform her of this option. The physician escaped liability, the court holding that *"the evidence overwhelmingly suggests that the mainstream medical community does not recognize or accept EDTA therapy as an alternative to a carotid endarterectomy in treating coronary blockages ... Opposition to EDTA therapy is based not only upon*

---

[283] News from the John A Burns School of Medicine University of Hawaii. Summer 2003.
[284] <u>Charrell v. Gonzales</u>, 673 N.Y.S. 2d 685 (1998).

the lack of objective evidence that the treatment is effective, but also upon evidence that the treatment may be dangerous."[285]

The allopathic physician, however, has the duty to keep up to date with therapeutic developments, including certain CAM practices. For example, a 1997 NIH consensus statement supported acupuncture as a legitimate therapy with proven efficacy for adult postoperative and chemotherapy-induced nausea and vomiting.[286] Paradoxically, as more scientific data about the efficacy of CAM treatment modalities becomes available, physicians may find themselves liable for failure to incorporate CAM options when informing their patients of treatment alternatives.

DEFENSES

There are several legal defenses for a physician's integrating, utilizing or supporting CAM therapies. One possible defense is to assert the 'respectable minority' standard of care, if it can shown that a respectable minority in the medical community also accepts the treatment in question. Courts differ as to what constitutes a respectable minority and there have been no cases to date that have raised this defense.

A second defense is assumption of risk. In *Schneider v. Revici*,[287] a physician delivered nutritional (selenium and dietary restrictions) and other nonsurgical treatment for breast cancer after the patient refused conventional treatments offered by other physicians. The patient signed a detailed consent form releasing the physician from liability and acknowledging that the defendant's treatments lacked FDA approval and that no results could be guaranteed. The cancer spread and patient sued for common law fraud, medical malpractice and lack of informed consent. The Court of Appeals held that assumption of risk is a complete defense to malpractice. The same court also held in another case that a patient's failure to sign a consent form did not preclude the jury from considering the assumption of risk defense, as consent may be written or verbal.[288]

---

[285] *Moore v. Baker*, 989 F.2d 1129 (11th Cir., Ga. 1993).
[286] NIH Consensus Development Panel on Acupuncture. *JAMA* 1998; 280:1518–24.
[287] *Schneider v. Revici*, 817 F. 2d 987 (2nd Cir., N.Y. 1987).
[288] *Boyle v. Revici*, 961 F.2d 1060 (2nd Cir., N.Y. 1992).

A third defense is to plead clinical innovation. An example of this is the off-label or non FDA-approved use of prescription drugs. Here, there is deviation from standard treatment in an attempt to alleviate a desperate situation, e.g., the patient is terminal or has failed conventional therapy.

Even if it's the patient's choice, physicians must still exercise due care when implementing therapy. In *Gonzalez v. New York State Department of Health*, Dr. Gonzales was charged with gross negligence and incompetence after he used nutritional therapies to treat six patients with incurable cancer who had failed or rejected conventional treatment. The hearing committee found that he missed signs of disease progression, and failed to perform adequate assessments, testing and follow-up evaluations. The court held that a patient's consent to or even insistence upon a certain treatment does not relieve the physician from the obligation of treating the patient with the usual standard of care.[289]

RISK MANAGEMENT

When discussing alternative therapy with a patient, the physician should first fully inform the patient about conventional treatments and their limitations. Next, the physician should explain why the novel, rather than the recognized conventional therapy, is being considered.[290] Finally, whether the physician intends to carry out CAM therapy, or refer to another practitioner, the patient must be warned about the potential risks associated with such therapy.

In order to guard against malpractice liability, Cohen and Eisenberg have recommended an approach to stratify risk depending on safety and efficacy. When evidence supports both safety and efficacy, the CAM therapy should gain general acceptance and become the standard of care. Where safety and/or efficacy are less well established, the physician should be much more guarded in offering the treatment. They should discourage patients from pursuing dangerous treatments such as injections of unapproved substances and pay close attention to known herb-drug interactions. Two examples are St John's wort interacting with oral contraceptives, chemotherapy agents and

---

[289] *Gonzales v. NYS DOH*, 232 A.D.2d 886 (1996).
[290] Court Recognizes Limits on Disclosure of Alternative Treatment. Hospital Law Newsletter 1994; 11:4–5.

immunosuppressants; and ginko biloba's effects on anti-clotting medications. The physician must also routinely inquire about herbal and home remedies when obtaining a medication history. If a patient insists on CAM treatments despite warnings, document the discussion carefully, including disclosure of evidence regarding potential dangers and lack of efficacy.[291]

---

[291] Cohen, MH and Eisenberg, DM. Potential Physician Malpractice Liability Associated with Complementary and Integrative Medical Therapies. *Ann Int Med* 2002; 136:596–603.

SUMMARY POINTS

COMPLEMENTARY AND ALTERNATIVE MEDICINE

- Patients are major consumers of Complementary and Alternative Medicine (CAM), and doctors should specifically seek this history in all patients.

- CAM practitioners such as chiropractors and acupuncturists face malpractice liability if they fall below the standard of care of their respective professions.

- Medical doctors who venture into the practice of CAM must adhere to CAM standards in addition to their own medical standards.

- CAM lawsuits against the medical doctor frequently allege lack of informed consent.

- It is necessary to keep up to date given the public's fascination with CAM so as to knowledgeably advise patients of CAM's potential benefits and harm.

14

Products Liability

When a person is injured by a defective product, be it a car or a prescription drug, he or she can file a cause of action against the manufacturer or others who have placed that item into the stream of commerce. The law governing this general area of product-related injuries, broadly termed products liability, covers more than one specific tort, and the injured plaintiff can recover damages under a number of legal theories. The defendants are frequently rich corporations such as pharmaceutical, tobacco, automobile, and asbestos manufacturers. Essentially, the claim is that the product was defective and unsafe, and the injury suffered by the plaintiff was a foreseeable consequence.

Litigation in products liability is big-time business, especially where many plaintiffs are injured, banding together to file a class-action lawsuit. Damages may be in the multi-million dollar range. In the medical field, products liability represents a very important area of litigation, and it is not difficult to see why. Patients are treated with medications that can have serious, even fatal, complications. Medical devices such as pacemakers and operating room equipment may malfunction. And mass vaccinations do give rise, if rarely, to neurological sequelae and fatalities. One would naturally ask upon whom, if any, should liability fall? The manufacturer? The hospital? The doctor?

THE LAW OF PRODUCTS LIABILITY[292]

Generally speaking, there are three legal theories that can be advanced in a claim under products liability: Negligence, breach of warranty, and strict products liability.

---

[292] Some of the hypotheticals and discussion in this section are adapted from a set of audiotapes on Torts by Professor Steven Finz, Sum & Substance First Year Superset ©1998.

Negligence: This is governed by the four elements of duty, breach, causation, and damages. The court examines whether there has been a breach of duty of reasonable care. Additionally, it looks at foreseeability.

Breach of Warranty: The word 'warranty' is borrowed from the law of contracts, and is defined by statutes governed by the Uniform Commercial Code, which has been adopted to provide uniformity in commercial activity. Simply put, there is a promise or 'implied warranty of merchantability' that the product is fit for ordinary use.

Strict Products Liability: This is the favorite legal basis for recovery in products liability cases, although in a jurisdiction like Massachusetts, breach of warranty of fitness is considered the functional equivalent of strict liability. Under strict liability, there is no need to prove fault or warranty. The legal doctrine had its genesis in a California case. Mr. Greenman was the plaintiff who was injured when he used a power tool that was given to him as a gift. He sued the manufacturer, although there was no direct contract of warranty between him and the manufacturer as he did not make the purchase himself (so-called lack of privity). By using the notion of strict liability, the California Supreme Court went beyond the contract theory to find the manufacturer liable:

> *"The remedies of injured consumers ought not to be made to depend upon the intricacies of the law of sales.... To establish the manufacturer's liability, it was sufficient that plaintiff proved that he was injured... as a result of a defect in design and manufacture... [making it] unsafe for its intended use."*[293]

The strict liability approach has been adopted, in some cases with modifications, in virtually all jurisdictions. The theory holds that a professional supplier who sells a product that is both defective and unreasonably dangerous is strictly liable to foreseeable plaintiffs. 'Defective' is usually defined as product quality below the expectations of a reasonable consumer. The phrase 'unreasonably dangerous' means the risks outweigh the advantages. These elements are illustrated by the following typical law school examples: When

---

[293] *Greenman v. Yuba Power Products Inc.*, 377 P.2d 897 (Cal. 1963).

Products Liability 141

someone is cut by a sharp knife, this does not signal a defective or unreasonably dangerous product, as knives are meant to be sharp. On the other hand, a drinking glass that is manufactured with a sharp edge is a defective product (not the usual expectation of the consumer) that is unreasonably dangerous, i.e., the risk of cutting the lips outweighs the advantage of its use as a drinking container. Manufacturers of products that are unavoidably unsafe, however, would be exempt from liability (see *Hines* case under Medical Products Liability).

Strict liability is not about negligence or fault, but is about a social policy that shifts to the manufacturer the cost of compensating the injured consumer. The rationale is that the manufacturer is able to pass on the damages by increasing the price of its products to all of its customers. Detractors claim that this raises the cost of doing business, and reduces the manufacturer's competitive edge. In any case, the plaintiff must still show proximate cause and damages, and assumption of risk is still a valid defense.

The most dramatic example of products liability is the recent case against the tobacco company Philip Morris, hit with a $3 billion verdict. The plaintiff, 56-year-old Richard Boeken, developed metastatic lung cancer from a lifelong habit of smoking cigarettes (Marlboro brand, manufactured by Philip Morris). He alleged that he was misled by the defendant about the dangers of cigarettes. After deliberating for nine days, a 12-member jury in Los Angeles awarded him $5.5 million in compensatory damages and an additional $3 billion in punitive damages. A trial judge reduced the punitive damages to $100 million,[294] which was further reduced to $50 million. Seeking exoneration, Philip Morris intends to appeal to the U.S. Supreme Court after the California Supreme Court refused to hear the case.

In 1999, the same tobacco manufacturer was ordered to pay $51.5 million, reduced by the judge to $25 million, to a former smoker. That case is also under appeal. On the other hand, repeated suits have been unsuccessful against the manufacturer of Bendectin, an anti-nausea drug used in pregnancy. The drug was alleged to have caused birth defects.[295]

---

[294] Philip Morris Is Hit With $3 Billion Verdict. *The Wall Street Journal*, June 7, 2001, pp. A-3; *The Los Angeles Times*, August 12, 2005.
[295] Sanders, J. *Bendectin on Trial: A Study of Mass Tort Litigation*. University of Michigan Press, 1998.

MEDICAL PRODUCTS LIABILITY

There have been many court decisions regarding defective products used in a medical setting, but results have not been consistent. In cases of adverse reactions to medications or diagnostic agents, the courts may take the view that the hospital is not in the primary business of supplying drugs. Contrariwise, the hospital may be found liable because it sold the product, as in one case involving radiographic contrast material that led to patient injury.

Products liability litigation adds significantly to healthcare costs. Manufacturers of drugs and devices must plan for lawsuits that inevitably follow the entry of their products into the marketplace. These projected costs are factored into the price, which means ultimately the consumer pays more. It has been estimated that in 1990, some $10.8 billion, not counting legal and administration costs, were paid to claimants alleging medical products liability.[296] Recent data continue to be unsettling. In 2004, Jury Verdict Research reported that the median compensatory jury award for medical products liability cases in 1993–2002 was $1 million. The median award in breast implant cases exceeded this figure, at $1.5 million, with plaintiffs winning in 36% of the cases. Injuries from drugs/vaccines had an even higher median award of over $2 million, with plaintiffs prevailing in 67% of the cases.[297] Additionally, class action suits are a particular problem as damages sought are notoriously very high.

The most currently talked-about example of products liability involves pharmaceutical defendant Merck & Co., whose product, Vioxx, a nonsteroidal COX-2 anti-inflammatory drug, is alleged to cause serious cardiovascular complications. Plaintiff won the first case (*Ernst v. Merck*), with the Texas jury on August 19, 2005 awarding damages of over $250 million dollars.[298] Some 5,000 additional Vioxx lawsuits are said to be pending. A second case argued before a New Jersey jury was won by Merck.

Blood Transfusions: Jurisdictions differ as to whether blood is to be considered a product. A 1954 New York case deemed blood transfusion a service rather than a product, and therefore exempted it from products liability based

---

[296] Lobe, TE. *Medical Malpractice: A Physician's Guide*. McGraw Hill, Inc. 1995, p. 375.
[297] JVR News Release. Available at www.juryverdictresearch,com. Accessed July 4, 2004.
[298] Merck Loss Jolts Drug Giant, Industry. *The Wall Street Journal*, August 22, 2005, p. A1.

on breach of warranty or strict liability.[299] Other courts have reached different results and have chosen to impose liability. Several state legislatures have enacted statutes protecting blood banks and healthcare providers from liability relating to blood transfusions.

An example where strict liability failed is Hines v St. Joseph's Hospital. In this case, the plaintiff, Tommie Hines, received blood transfusions and contracted hepatitis in 1970. The defendants were Blood Services Inc., the supplier of the blood, and St. Joseph's Hospital, where the transfusions were ordered and given. In ruling for the defendants, the Court essentially held that the product, blood in this case, was not unavoidably unsafe, and therefore not unreasonably dangerous, and relied on the language of the Restatement (Second) of Torts, §402A (1965):

> "k. Unavoidably Unsafe Products. There are some products, which, in the present state of human knowledge, are quite incapable of being made safe for their intended and ordinary use.... Such a product, properly prepared, and accompanied by proper directions and warning, is not defective, nor is it unreasonably dangerous... The seller of such products, again with the qualification that they are properly prepared and marketed, and proper warning is given, where the situation calls for it, is not to be held to strict liability for unfortunate consequences attending their use, merely because he has undertaken to supply the public with an apparently useful and desirable product, attended with a known but apparently reasonable risk."

At the time of Hines' transfusion, no test could adequately detect the hepatitis virus in blood. Further, no process could destroy the virus without damaging the blood. The court reasoned that blood was a product that was quite incapable of being made entirely safe for its intended and ordinary use. Nonetheless, blood for transfusions was both a useful and desirable product and was at times absolutely essential to save life. No liability was found, as the risk was outweighed by this public benefit and was thus reasonable.[300]

---

[299] Perlmutter v. Beth David Hospital, 123 N.E.2d 792 (N.Y. 1954).
[300] Hines v. St. Joseph's Hospital, 527 P.2d 1075 (N. Mex. 1974).

A more recent case involved HIV-contaminated blood. In *Snyder v. American Association of Blood Banks*, the Supreme Court of New Jersey found the trade organization liable for failing to recommend routine HIV testing of the nation's blood supply in the early 1980s. The Court reasoned that the Association breached its duty of due care as it had projected itself as the guardian of blood safety and therefore ought to have recommended HIV testing even though the data were inconclusive at the time.[301]

Prescription Drugs and Medical Devices: Hospitals usually escape liability for adverse drug reactions, with courts tending to shield them against implied warranty (hospitals are not merchants) or strict liability. In a case where the dose of X-rays was at issue, the Illinois Supreme Court refused to allow strict liability consideration by ruling that the X-rays themselves were not a defective product.[302] Similar approaches govern medical devices. For example, a California court held the hospital a user rather than a supplier of a surgical needle that broke. Cases involving pacemakers and spinal implants have met with similarly favorable results. There are of course court decisions to the contrary, and these cases have involved such items as suturing-needles, jaw implants, even flammable hospital gowns. Federal medical device laws do preempt some but not all state-court damage actions under products liability.[303]

A dramatic medical product liability case was recently decided by a Texas jury, which awarded over $1 billion to the family of Cynthia Cappel-Coffey, a 41-year-old woman who died of primary pulmonary hypertension (PPH) after using the prescription diet drug Pondimin (Fenphen). The defendant was Wyeth, a large pharmaceutical company. The jury awarded $113 million in actual damages, and $900 million in punitive damages, making it the largest verdict for products liability involving a drug or medical device. Arguing that there was absolutely no basis in the record for the amounts awarded, and that the verdict was contrary to Texas' own statutory cap on punitive damages,

---

[301] *Snyder v. Am. Ass'n. of Blood Banks*, 676 A.2d 1036 (N. J. 1996).
[302] *Shivers v. Good Shepard Hospital*, 427 S.W.2d 104 (Tex. Civ. App. 1968).
[303] Medical Device Amendments of 1976, Federal Food, Drug, and Cosmetic Act, §510(k), as amended, 21 U.S.C.A. §360(k).

Wyeth has promised to appeal the verdict. In a press release, the company asserted:

> "Ms. Cappel-Coffey, who was morbidly obese... had a family history of cardiovascular disease... did not develop PPH symptoms until more than four years after she stopped using Pondimin. There are no studies that demonstrate an association between Pondimin use and an increased risk of PPH that long after cessation of use."[304]

'Learned Intermediary' Doctrine: Generally speaking, if a doctor fails to warn the patient of a medication risk, the patient has a claim against the doctor but not the drug manufacturer. This is termed the 'learned intermediary' doctrine. The doctrine is equally applicable to medical devices and bodily implants. The justification is that the manufacturer can reasonably rely on the treating doctor to warn of the adverse effects which are well catalogued in the PDR. Additionally, pharmaceutical representatives are supposed to warn of a drug's side effects whenever they call on the doctor, who is then expected to use his or her professional judgment to adequately warn the patient. It is also not feasible for the manufacturer to directly warn every patient, without usurping the doctor-patient relationship. There are, however, exceptions as in mass inoculations and birth control drugs where the manufacturer has the obligation to directly warn the public.

In addition to the failure to warn, physicians are at risk for prescribing the wrong medication, using the wrong dose, administering the drug via the wrong route, or failing to recognize a foreseeable drug-drug interaction or an allergic reaction. 'Off-label' drug use also puts the physician at risk and requires an awareness of contraindications and careful discussions with the patient.

The 'learned intermediary' doctrine may not always immunize drug manufacturers from a lawsuit. The pharmaceutical industry relies upon the prescribing physician to adequately warn patients of the side-effects of prescription drugs, hoping thereby to escape liability for adverse consequences. However, where the manufacturer has reason to know that the drug will reach the

---

[304] Available at www.wyeth.com. Accessed September 25, 2004.

consumer without the intervention of a physician, it must take reasonable action to warn the consumer directly. Of present-day relevance is the massive direct-to-consumer advertising that pharmaceutical companies are conducting over the Internet and in the mass media. This clearly places drug manufacturers at increased risk for liability.[305]

Even if printed warnings comply with requirements of the Food and Drug Administration (FDA), the manufacturer may still face liability if its sales personnel engage in high-pressure promotion of the drug and minimize or ignore the risks. Where the manufacturer is unaware of the danger regarding an FDA-approved drug, the courts have generally denied liability. However, the defendant will be liable if it is negligent in testing the product and fails to find out that it's unsafe. A striking example is the MER/29 litigation. The drug was developed and found to be effective for reducing serum cholesterol, but it caused cataracts and was subsequently withdrawn from the market. The earliest cases held the seller not liable, on the ground that the danger was undiscoverable. Later, attorneys uncovered the fact that the manufacturer was aware that its tests were not conclusive or satisfactory, and that it had manipulated the test results in reports to the FDA to obtain a permit to market the drug. On this basis, recovery was allowed for negligence.

---

[305] Mcintire, T. Legal and Quality of Patient Care Issues Arising From Direct-to-Consumer Pharmaceutical Sales. 33 *U. Mem. L. Rev.* 105 (2002).

SUMMARY POINTS

PRODUCTS LIABILITY

- The manufacturer's liability is based on the legal theories of negligence, contractual warranty, and strict liability.

- Strict liability looks not at fault but at whether there is a defect that makes the product unreasonably dangerous, and whether the injury is foreseeable.

- Products liability litigation significantly adds to healthcare costs.

- As the 'learned intermediary,' the medical practitioner is typically liable for adverse results arising from the use of medical products or devices.

- Failure to warn and to recognize a potential allergic reaction are common grounds for litigation.

15

Institutional and Managed Care Liability

When a negligent act occurs in a healthcare setting, the question arises as to whether parties other than the tortfeasor, such as the employer or the institution, may also be legally responsible for harm to the patient. This form of indirect fault is termed vicarious liability. Institutions may also be directly liable for their own negligence.

INSTITUTIONAL LIABILITY

In the distant past, hospitals were charitable institutions that provided free treatment to the indigent poor. Most of them were faith-based or state-sponsored, and they enjoyed immunity from lawsuits. In more modern times, however, hospitals have evolved into large institutions, frequently for-profit, and are recognized as places offering diagnostic studies and medical and surgical treatment. High standards are expected of these centers and the immunity from lawsuits they once enjoyed was gradually lost. Today, hospitals, like the doctors and nurses who work there, can be sued for wrongful conduct that leads to patient injury. The liabilities can be direct or indirect.

Direct Liability: In addition to direct liability, hospitals may be held vicariously liable (i.e., indirectly liable) for the negligent acts of its doctors and staff. Examples are: short-staffing, negligent nurse-hiring practices, or negligent credentialing of doctors. A hospital can also be sued for injuries that occur on its premises (called premise liability), e.g., a patient or visitor slipping on a wet slippery floor.

Then there is the concept of corporate liability. In the landmark case of *Darling v. Charleston Community Memorial Hospital*, the court held that the hospital was liable for the negligence of a non-employee emergency-room doctor, a contract physician, whose substandard care ultimately led to the amputation of an athlete's leg.[306] The court reasoned that a hospital is required to ensure acceptable patient-safety procedures and competency of its entire medical staff. In another notorious case, the hospital was likewise held liable for the negligence caused by a neurosurgeon who performed unnecessary surgery with resulting harm to patients.[307]

Another legal theory for direct hospital liability is based on the concept of non-delegable duty. Certain duties may be held to be non-delegable, i.e., they cannot be delegated by the hospital to others, including doctors, and the hospital remains responsible for them. The example commonly given is the hospital's emergency services, an inherent function of the hospital. In an Alaska case, the court ruled that such emergency duties were non-delegable, and held the hospital liable even though its emergency room physicians, who were independent contractors and not hospital employees, committed the tortious act.

Of recent interest are staff duties in hospital emergency departments in screening and stabilizing patients. This is an area of rising liability. The federal law called EMTALA (Emergency Medical Treatment and Active Labor Act) tightly regulates the activities of a hospital emergency department, and provides for severe penalties in the event of a violation. EMTALA's main purpose is to prevent the dumping of poor, non-paying patients by the facility. It requires hospitals to assess the patient's condition and to stabilize the patient before transfer to another facility.

Finally, hospitals can be held liable based not on negligence, but on strict liability principles. Strict liability concerns typically arise from abnormally dangerous activities such as blasting, fumigation and other inherently dangerous activity. Some types of hospital activities such as chemotherapy or radiation therapy, may be covered under this description. Or the hospital may be construed as functioning as a seller of defective products that are unreasonably dangerous. Strict liability may attach if a patient is harmed as a result (see previous chapter on Products Liability).

---

[306] *Darling v. Charleston Community Memorial Hospital*, 211 N.E.2d 253 (Ill. 1965).
[307] *Gonzales v. Nork*, Suit No. 22856, Super. Ct. Cal. (1973).

Vicarious Liability: In addition to direct liability, how else can hospitals be held liable for the negligent acts of its doctors and staff? Court decisions have generally used the employer-employee or the agency principle to hold hospitals vicariously liable. Where there is an employer-employee relationship, e.g., nurses and some doctors hired by the hospital, *respondeat superior* is the basis for liability. *Respondeat superior* means: 'Let the master answer.' So long as an employee (or the old term 'servant') carries out the negligent act during and within the scope of the employment, the employer will be held vicariously, i.e., indirectly, liable. The idea behind this rule is to ensure that the master or employer, as supervisor, will enforce the proper work standards so as to avoid risk of harm. Thus, liability flows back to the master for the negligence of the servant. In the hospital setting, nurses and other employed hospital workers are covered by the *respondeat superior* doctrine. The test is whether the act occurred during and within the scope of employment, and whether the risk of harm was foreseeable.

However, where the negligent actor such as a doctor, is an independent contractor rather than an employee, *respondeat* will not apply. An institution usually does not exercise substantial control over the actions of independent contractors, whereas the opposite is true for its employees. Most doctors who work in private hospitals are independent contractors, as they do not draw a hospital salary, nor are their work hours and work duties controlled or defined by the hospital.

Depending on the facts, some courts have concluded that there may be an agency relationship between the doctor and the hospital. Agency may be established if there is some degree of control, even if minimal, that is exerted on the doctor by the hospital. The relationship may be construed as an ostensible agency, where there is some representation that the doctor works for the hospital. When the patient relies on this relationship in seeking his or her treatment, it is called agency by estoppel. Under any of these formulations, the plaintiff may include the hospital as a co-defendant in a malpractice suit.

Thus, invoking vicarious liability begins with a determination as to whether the tortfeasor is an employee. Then the court will look at whether there is an agency relationship. If both are absent, there will be no vicarious liability. The issue may not always be straightforward, as a recent Florida case illustrates: The ship's doctor aboard a Carnival cruise ship failed to diagnose acute appendicitis in a 14-year-old girl despite several days of abdominal symptoms. The patient ruptured her appendix, and this resulted in sterility. The parents sued the cruise line as a codefendant. Although the doctor's contract stated that he was an independent contractor, the court reasoned that in a claim based on agency, it is the right of control rather than actual control itself that matters. It therefore held that: *"(1) for purposes of fulfilling cruise line's duty to exercise reasonable care, ship's doctor is an agent of cruise line whose negligence should be imputed to cruise line, regardless of contractual status ascribed to doctor; and (2) to the extent cruise ticket sought to limit cruise line's liability for negligence of doctor, it was invalid."*[308]

Liability of Doctors for Their Employees: The doctor is responsible not only for his or her own professional conduct, but also for that of staff members. Vicarious liability may be imposed on the doctor for the negligent acts of office assistants, nurses, technicians and other staff under the legal theory of *respondeat superior*. This may also be true where a doctor hires a *locum* or a junior doctor as an employee. The wrongful acts must have been committed in the course of the employment.

Another variation is that of a surgeon who works with nurses and other hospital employees ('borrowed servants'). The surgeon is deemed the 'captain of the ship,' and will be held liable for harm resulting from the negligent acts of the assistants. Today, the 'borrowed servant' is no longer an important legal rule, because hospitals no longer enjoy tort immunity, and there is no need to reach for the deep pocket of the surgeon.

Doctors may also be liable for the negligent acts of house-staff members under their supervision (see Chapter 16: Medical Trainees).

MANAGED CARE LIABILITY

Within the past two decades, managed care, covering four out of five working Americans, has emerged as the choice model for healthcare delivery. Managed care is defined by the following features: (1) It is an integrated prepaid insurance system for the delivery of medical care; (2) It contracts with selected physicians and hospitals to provide these medical services; (3) It enrolls workers, usually from large corporations and businesses, as members of the managed care 'plan' for a predetermined premium; and (4) It attempts

---

[308] *Carlisle v. Carnival Corporation, et al.*, 864 So.2d 1 (Fl. 2003).

to control cost by modifying the behavior of doctors who have signed on as providers in the plan. The term managed care organization (MCO) is used synonymously with the term health maintenance organization (HMO).

With managed care, doctors become 'double agents,' with loyalties not only to their patients, but also to the plan that pays their salaries and provides incentives for productivity and cost-containment. This poses a potential conflict of interests. A cost-sensitive MCO, the only kind that will survive, has a powerful incentive to do less, which may or may not undermine the quality of care, whereas the doctor-patient relationship, rooted in trust, requires the doctor to do everything in the patient's best interests. Doctors also resent being controlled by non-medical administrators regarding the practice of clinical medicine. Utilization review, gag rules, incentives and disincentives, and gatekeeper techniques are among the frequent complaints. Doctors are therefore outraged when they discover abuses, as in the rare instance where the money 'saved' was used to line the pockets of administrators, directors, and shareholders of MCOs, some of whose CEOs may command annual salaries in the millions.

Patients too can dislike managed care, which is perceived as limiting their choice of caregivers, reducing face-to-face time with doctors, unreasonably shortening the length of hospital stays, and restricting expensive but necessary tests and treatments. According to *The Economist*, a British-based weekly magazine, managed care "*has taken over from communism as the bogeyman that unites the country in fear and loathing,*" with only 15% of Americans having a great deal of confidence in them, which is substantially less than confidence levels for electricity companies (28%) or public schools (38%).[309]

However, some of this scorn is misdirected. Keeping healthcare costs affordable is a worthy societal goal, and many MCOs such as the Kaiser system do a credible job in reining in profligate expenses without sacrificing quality of care.

Liability of MCOs: An MCO may be held vicariously liable under the doctrine of ostensible agency, i.e., liability is based on whether the patient looks to the organization rather than the individual physician for care, and whether the MCO holds out the physician as its employee. The theory was tested in *Boyd v. Albert Einstein Medical Center*, a case in which a breast biopsy

---

[309] *The Economist*, June 23, 2001, p. 33.

caused a hemothorax, and there was failure to diagnose a concurrent acute myocardial infarct. The plaintiff alleged that the physicians were negligent and therefore the Medical Center was vicariously liable. The court held that it was a question of fact (i.e., the jury gets to decide) whether the participating physicians were ostensible agents of the MCO, and that the case was therefore improperly dismissed by the judge at an earlier stage in favor of the defendant-MCO.[310]

Undue deference to an insurer can put doctors at legal risk. The classic case in point is *Wickline v. State*, which raised a liability issue for the state of California under its medical insurance plan, Medi-Cal. The patient, Lois Wickline, suffered from Lerische's syndrome (obstruction of the terminal aorta with arterial insufficiency in the groin and legs). The treating doctors discharged the patient prematurely because their initial request for an eight-day stay was rejected by Medi-Cal, which favored a shorter four-day hospital stay. As a result of her premature discharge, the patient developed gangrene and lost her right leg. The issue was whether Medi-Cal was liable, or whether the doctors themselves should have filed another request for extension after the initial four-day period had lapsed when in their judgment, an extension was medically necessary. The court ruled for the State of California, holding that: "*Medi-Cal did not override the medical judgment of Wickline's treating physicians at the time of her discharge. It was given no opportunity to do so. Therefore, there can be no viable cause of action against it for the consequences of that discharge decision.*"[311] Lesson: Doctors should advocate for the patient and must exhaust all avenues (with detailed documentation) before surrendering to a bad decision by insurance payers.

Failure to make appropriate referrals in the name of cost savings can also get the provider into trouble. It was the basis for the large $89 million verdict against Health Net, California's second largest HMO, in 1994. In that case, the HMO refused to authorize a bone marrow transplant for a woman with metastatic breast cancer.[312]

ERISA: A peculiar aspect of labor law has kept many lawsuits against HMOs out of state courts. A federal statute known as ERISA[313] governs pension plans and employee benefits that usually include healthcare benefits. Under

---

[310] *Boyd v. Albert Einstein Medical Center*, 547 A.2d 1229 (Penn. 1988).
[311] *Wickline v. State*, 192 Cal.App.3d 1630 (1986).
[312] Gorney, M. Liability Pitfalls in Managed Care. *The Doctor's Advocate*, 3rd Quarter, 1996.
[313] Employee Retirement and Income Security Act of 1974 (ERISA).

ERISA, a beneficiary can sue a benefit plan, a benefit plan manager, or other fiduciary to recover benefits, i.e., the cost of medical treatment due to him or her, but cannot collect personal compensatory or punitive damages for any injuries that may have resulted. On the other hand, if the beneficiary can successfully sue in state court on an allegation of medical malpractice or other tort action, then the monetary damages, both compensatory and punitive, are substantially higher.

ERISA supersedes state laws on matters relating to any employee benefit plan that is covered by ERISA.[314] Thus, suits alleging damages from malpractice actions are foreclosed under the terms of ERISA. This quirk in the law therefore places plaintiffs in an ERISA-qualified health plan in a disadvantaged position compared to those in a non-ERISA health plan such as a government-sponsored or private health plan.

The controversy surrounding ERISA and the right to sue for medical malpractice reached a legal zenith in *Pegram v. Herdrich*, a case that was recently decided by the U.S. Supreme Court.[315] Cynthia Herdrich sued her managed care plan, Carle HMO, for the failure of its doctor, Dr. Pegram, to timely diagnose and treat her appendicitis. The patient's diagnosis was delayed after Dr. Pegram found a 6 × 8 cm inflamed mass in the abdomen because she had to wait eight additional days for a confirmatory ultrasound procedure at a participating facility 50 miles away. A closer facility would have promptly carried out the test, but it was not chosen because it was not affiliated with the HMO. With the delay, her appendix ruptured. The lower courts decided in favor of the plaintiff, i.e., allowed her to sue the HMO, but upon appeal, the U.S. Supreme Court refused to hold the HMO liable. The Supreme Court held that ERISA's 'fiduciary duty' provision did not apply to financial arrangements between plans and physicians that reward the latter for withholding treatment. Concerned about "*precipitate upheaval*" and specifically noting Congress's creation of the HMO model in the first place, the Court's decision in effect served to immunize ERISA-approved health plans from liability for utilization management gone awry.

---

[314] ERISA, 29 U.S.C.A. §1144(a).
[315] *Pegram v. Herdrich*, 120 S. Ct. 2143 (2000).

On June 21, 2004, the Supreme Court handed down its latest opinion on HMO liability.[316] In two related and consolidated cases, the Court again held that patients do not have the right to sue their HMOs in state courts over ERISA-sponsored employee benefits including healthcare benefits. The plaintiffs had sued their HMOs (Aetna Health in one case, Cigna Healthcare of Texas in the other) in state court under the Texas Health Care Liability Act for injuries suffered due to the administrators' refusal to cover treatment recommended by their physicians. In the first case, the patient bled from the use of anti-inflammatory drugs that replaced the doctor's choice of a safer drug. The second case dealt with a 'medically necessary' one-day hospital stay for a hysterectomy. This was against the doctor's recommendation, and the patient suffered complications as a result of the early discharge.

---

[316] *Aetna Health Inc., v. Davila,* and *Cigna Healthcare of Texas v. Calad,* 124 S. Ct. 2488 (2004).

SUMMARY POINTS

INSTITUTIONAL AND MANAGED CARE LIABILITY

- Tort liability can be direct or indirect. The latter is termed vicarious liability.

- Courts have used legal theories of corporate liability, non-delegable duties and inherent function to directly implicate the institution.

- Institutions are vicariously liable for their employees under the doctrine of *respondeat superior.* However, they are usually not liable for the negligence of independent contractors, unless they are deemed to be agents of the institution or the duty is considered non-delegable.

- Doctors are likewise vicariously liable for the negligence of their employees during and within the scope of employment.

- Managed care organizations have been shielded from tort action because of a quirk in the ERISA law.

16

Medical Trainees[317]

Medical trainees, i.e., fellows, residents and students, are regularly joined with supervising physicians (typically medical school faculty or community practitioners) and the hospital as co-defendants in a malpractice lawsuit. This is said to occur in nearly 25% of cases where trainees are part of the healthcare team.[318] Trainees are responsible for care to patients under their charge, but are required to be properly supervised. The supervising physician continues to be liable if harm should result from the trainee's negligence. In a recent case, a fourth year medical resident inserted a trochar into the abdomen of a patient under the supervision of a faculty physician. The patient died, and the faculty member was held vicariously liable since the hospital by-laws required that *"only a licensed physician with clinical privileges shall be directly responsible for a patient's diagnosis and treatment."*[319] Although still acquiring the skills towards certification, trainees are nonetheless legally responsible for their actions. But does the law demand the same standard of care as it would a fully qualified attending physician? The answer is a surprising yes.

Both the frequency and severity of malpractice suits involving trainees are increasing. According to one source, the number of suits has doubled between 1975 and 1979, and doubled again between 1985 and 1989. The National Practitioner Data Bank (NPDB) lists the names of 1,331 residents as of the year 2,002.[320] About 1% of all claims reported to NPDB include

---

[317] Dr. Laura Rogers, Internal Medicine resident, assisted in the research for this chapter during a senior medical student elective.
[318] Kachalia, A and Studdert, DM. Professional Liability Issues in Graduate Medical Education. *JAMA* 2004; 292:1051–6.
[319] *Brown v. Flowe*, 496 S.E.2d 830 (N.C.App. 1998).
[320] Helms, L and Helms, C. Forty Years of Litigation Involving Residents and Their Training; II. Malpractice Issues. *Acad Med* 1991; 66:718–25; National Practitioner Data Bank 2002 Annual Report, pp. 26–7.

trainees as defendants.[321] It is believed that the increased number of residents in training, the increased complexity of the resident's practice environment, and the overall litigation climate play contributory roles.

LIABILITY OF TRAINEES

Standard of Care: In the past, courts ruled in favor of a dual standard of conduct for trainees and licensed practitioners. That is, trainees were held to a lower standard of care whereas certified physicians had to meet the standard ordinarily expected of physicians under similar circumstances. This dual standard was articulated in Rush v. Akron General Hospital, which involved a medical intern who treated a patient who had fallen through a glass door. The patient suffered several lacerations to his shoulder, which the intern treated. However, when two remaining pieces of glass were later discovered in the plaintiff's shoulder, he sued the intern for failing to adequately examine and treat the wounds. The court dismissed the claims against the intern, finding that he had practiced with the skill and care of his peers of similar training: *"It would be unreasonable to exact from an intern, doing emergency work in a hospital, that high degree of skill which is impliedly possessed by a physician and surgeon in the general practice of his profession, with an extensive and constant practice in hospitals and the community."*[322]

This dual standard began to erode when several court decisions favored a single standard of conduct for residents and fully qualified practitioners. Pratt v. Stein involved a second year orthopedic resident who treated a postoperative patient for an infection. The resident allegedly administered a toxic dose of neomycin, which resulted in deafness. The lower court held that the resident should be judged by a lower standard than that expected of a fully trained specialist. However, on appeal, the court reversed, and held that the resident should be judged by the same standard as a specialist when he practiced in that area. The court reasoned that *"... a resident should be held to the standard of a specialist when the resident is acting within his field of specialty. In our estimation, this is a sound conclusion. A resident is already a physician who has chosen to specialize, and thus possesses a higher degree of knowledge and skill in the chosen specialty than does the non-specialist."*[323]

---

[321] NPDB Public Use Data File. Available at: www.npdb-hipdb.com. Accessed October 16, 2004.
[322] Rush v. Akron General Hospital, 171 N.E.2d 378 (Ohio Ct.App. 1987).
[323] Pratt v. Stein, 444 A.2d 674 (Pa. Super. 1980).

Courts continue to demand a unitary standard regardless of the doctor's training status. This is in part because trainees are expected to perform under careful and direct supervision. Some academics have asserted that "*if one is holding oneself up as providing a certain level or type of care, it becomes the duty of that individual — whether dentist or nurse, student, intern, or resident — to provide care in accordance with the applicable standard of conduct for that profession.*"[324]

This trend of a single uniform standard has its exceptions. Occasionally, the courts have supported intermediate standards for residents that more appropriately reflect their level of training. Pennsylvania is such a jurisdiction. Resident skill improves over time, and depending on how long the trainee has been in the residency program, he or she may be held to some standard that corresponds to that unique level of knowledge and experience. In *Jistarri v. Nappi*, an orthopedic resident allegedly applied a cast with insufficient padding to the broken wrist of a patient. The plaintiff claimed this led to soft tissue infection with *Staph. aureus*, with complicating septicemia, staphylococcal endocarditis and eventual death. The court held that the resident's standard of care should be:

> "*... higher than that for general practitioners but less than that for fully trained orthopedic specialists ... To require a resident to meet the same standard of care as fully trained specialist would be unrealistic. A resident may have had only days or weeks of training in the specialized residency program; a specialist, on the other hand, will have completed the residency program and may also have had years of experience in the specialized field. If we were to require the resident to exercise the same degree of skill and training as the specialist, we would, in effect, be requiring the resident to do the impossible.*"[325]

Duty to Treat: Trainees are not at risk only for substandard conduct. They may face issues regarding whom they are obligated to treat. Duty to care for a patient arises from the doctor-patient relationship and legal duty to treat does

---

[324] Butters, J and Strope, J. Legal Standards of Conduct for Students and Residents: Implications for Health Professions Educators. *Acad Med* 1996; 71:583–90.
[325] *Jistarri v. Nappi*, 549 A.2d 210 (Pa. Super. 1988).

not apply to strangers. However, trainees in the hospital environment are usually obligated to care for all in-patients during an emergency, whether they be the resident's assigned patients or not. For example, during cardiopulmonary arrest, it is usually the resident's job to direct the resuscitation of the patient.

Good Samaritan statutes protect against negligence when the rescuer comes to the aid of strangers, but most statutes carve out an exception for rescue in a healthcare institution where the law accords no special immunity.[326]

Representation and Informed Consent: Medical trainees have a legal duty to represent themselves to patients and their families according to their educational status. Although medical students are often introduced as or assumed to be doctors, the patient should be made aware of their level of training before accepting them as healthcare providers. The supervising physician is responsible for ensuring that the status of each member of the medical team is unambiguously understood by the patient. The patient's right to decline treatment by the medical trainee should also be made clear. These issues were raised by the Secretary of Health, Education, and Welfare's Commission on Medical Malpractice in 1973, which urged that each patient admitted to a teaching hospital be made aware of the educational functions of that hospital. Failure to disclose the educational status of caregivers can result in torts of lack of informed consent, misrepresentation, fraud, deceit, invasion of privacy, and breach of confidentiality. In several states it is a misdemeanor to use the professional title of 'doctor' without legal authorization.[327]

Another issue is informed consent. Depending on the situation, this duty to obtain permission from the patient may fall, at least initially, on the trainee. Failure to obtain informed consent results in liability for the tort of battery and/or negligence.

Communication: Although poor communication skills are not legal elements of malpractice claims, a poor doctor-patient relationship may prompt patients to sue should an adverse outcome occur. Beckman et al. found that communication problems were part of over 70% of all malpractice suits.[328] Positive communication skills such as eye contact, smiling, active listening, and using

---

[326] See, for example, Hawaii's Good Samaritan statute at HRS §663-1.5.
[327] Kapp, M. Legal Implications of Clinical Supervision of Medical Students and Residents. *J Med Ed* 1983; 58:293–9.
[328] Beckman, HB et al. The Doctor-Patient Relationship and Malpractice. *Arch Int Med* 1994; 154:1365–70.

a friendly tone of voice will foster trust and rapport. Providing an honest and empathetic disclosure of mistakes when they occur can also minimize litigation.

Malpractice Insurance for Trainees: All trainees are covered by malpractice insurance of one type or another. Medical students are usually insured through the school's malpractice policy, which is typically a comprehensive umbrella policy that covers its students and faculty. Liability coverage is typically $1 million per incident, but can vary widely from a low of $50,000 to as high as $5 million.

ACGME (Accreditation Council for Graduate Medical Education) requires all medical residents to be insured through their graduate medical education program. The regulation reads, in part: *"Residents in the GME must be provided with professional liability coverage for the duration of training. Such coverage must provide legal defense and protection against awards from claims reported or filed after the completion of GME if the alleged acts or omissions of the residents are within the scope of the education program. The coverage to be provided should be consistent with the institution's coverage for other medical/professional practitioners. Each institution must provide current residents and applicants for residency with the details of the institution's professional liability coverage for residents."* [329] In other words, trainee insurance provides 'tail coverage,' which covers claims against events occurring during the training period, even if they are filed after residency is over and the resident is no longer insured with that carrier.

Note that coverage is provided only when the resident is practicing within the scope of the training program. If a resident decides to moonlight or be employed outside of the training program, the resident should either purchase separate professional liability insurance or have the employer provide such insurance.[330]

LIABILITY OF SUPERVISORS

Vicarious Liability: Trainee supervision by an attending physician serves two purposes: it allows them to improve their skills under the guidance

---

[329] American Medical Association, Graduate Medical Education Directory 2003–4, p. C.5:15.
[330] Berlin, L. Malpractice Issues in Radiology: Liability of the Moonlighting Resident. *A J Rad* 1998; 171:565–7.

of experienced physicians, and it ensures that patients will receive appropriate care. Attending physicians are directly liable for their own negligent care of patients, as well as vicariously liable for the resident's actions because of their supervisory role, although some authors have considered failure to supervise a form of direct rather than vicarious liability.[331] Trautlein et al. found that in 200 consecutive malpractice cases involving residents working in the emergency department, litigation resulted from "*housestaff officers apparently functioning in a non-supervised capacity, or residents on rotation from specialty training or moonlighting in an unsupervised capacity.*"[332]

The attending physician has historically been viewed as the 'captain of the ship' and held liable for all actions taken by the assistants he/she has control over. The resident can be viewed legally as a 'borrowed servant' who has been 'loaned' to the physician in charge of the case.[333] In *Rockwell v. Stone*, an anesthesiology resident missed the patient's vein when he tried to inject sodium thiopental intravenously. The intra-arterial or extravasated injection (which one happened was unclear) of this induction agent led to arterial vasospasm and thrombosis, irreversibly compromising the blood supply to the patient's arm and eventually necessitating amputation. The chief of anesthesiology, who was the supervisor of the resident, was found vicariously liable for the resident's negligence.[334]

In another case, an Ob/Gyn resident performed a tubal ligation, but the patient subsequently became pregnant and underwent a therapeutic abortion, followed by a hysterectomy. The court decided that: "*Even though the surgical procedure was actually performed by a resident, the attending physician and hospital were under a duty to see that it was performed properly. It is their skill and training as specialists which fits them for that task, and their advanced learning which enables them to judge the competency of the resident's performance.*"[335]

---

[331] Kachalia, A and Studdert, DM. Professional Liability Issues in Graduate Medical Education. *JAMA* 2004; 292:1051–6.
[332] Trautlein, JT et al. Malpractice in the Emergency Department — Review of 200 Cases. *Ann Emerg Med* 1984; 13:709–11.
[333] Holder, A. Liability for Resident's Negligence. *JAMA* 1970; 213:181–2.
[334] *Rockwell v. Stone*, 173 A.2d 48 (Pa. Super. 1961).
[335] *McCullough v. Hutzel Hospital*, 276 N.W.2d 569 (Mich. App. 1979).

What about the liability of the on-call supervising attending? Courts are divided on this issue. The on-call physician customarily takes calls from home, and may not have previously met the patient, and therefore no doctor-patient relationship has formed. Although there is a duty to supervise the trainee(s), the on-call status alone may not be enough to create a doctor-patient relationship. Such were the findings of a court which dismissed a negligence claim for failure to supervise two emergency room residents in the treatment of a young girl who died with undiagnosed chicken pox pneumonia.[336] On the other hand, an on-call agreement was sufficient for another court to impose a doctor-patient relationship upon the supervising attending, with concomitant duty of due care to the patient. That case involved mismanagement of labor that resulted in serious neurologic injury to the newborn.[337]

A supervisor may not always be found liable, especially where the trainee was performing tasks that he or she was reasonably expected to know in the absence of the supervising physician. In *Richardson v. Denneen*, the surgical attending asked the resident to finish suturing and dressing the incision before leaving the operating room. The resident applied phenol instead of alcohol to the skin, injuring it. In assigning the liability, the court found that it was proper practice for an attending to leave the operating room while the resident sutured the skin — a simple task. Accordingly, the attending surgeon escaped liability.[338] In another case, a neurosurgery resident failed to respond to calls from the recovery-room nurses who had noticed that the patient was displaying decreased movement of the extremities. As a result of the delay, the patient expired from a blood clot that had compressed the spinal cord. The resident was found liable for negligent care but the supervising neurosurgeon was not, because "*there was evidence that neurosurgeon taught students to respond quickly when there was decreased movement post surgery.*"[339]

The hospital may be vicariously liable for a resident's negligence through the doctrine of *respondeat superior* if the resident is deemed an employee of the hospital. Whether residents are considered employees or students with respect to the hospital is an issue of some debate. The National Labor Relations Board, the Internal Revenue Service, and state courts are at odds over the definition. Although trainees clearly have an educational purpose in their

---

[336] *Prosise v. Foster*, 544 S.E.2d 331 (Va. 2001).
[337] *Lownsbury v. VanBuren*, 762 N.E.2d 354 (Ohio 2001).
[338] *Richardson v. Denneen*, 82 N.Y.S.2d 623 (N.Y. Super. 1947).
[339] *Parmelee v. Kline*, 579 So.2d 1008 (La. App. 1991).

work, the courts have generally ruled that they are hospital employees for legal purposes of ascertaining vicarious liability.

Educational Negligence: A related issue is whether there has been appropriate trainee education. Negligent acts of students and residents could potentially involve program directors and teaching faculty. Those who guide morning report, attending rounds, noon conferences and other educational activities but who are uninvolved in direct patient care may still be legally responsible for information they provide to trainees. Although malpractice suits have been filed against educational entities, universities, and residency directors, they are largely unsuccessful due to the difficulty in proving faulty education and causation, as well as the probability of contributory negligence on the part of the trainee.[340]

Swidryk v. St. Michael's Medical Center was a rare lawsuit of this type. Dr. Swidryk was in his third week of obstetrical training when he delivered a patient who developed birth difficulties and brain damage. When he was sued for malpractice, Dr. Swidryk in turn sued the Director of Medical Education, alleging that the director's failure to educate and supervise adequately was the proximate cause of his negligent care. The New Jersey Appellate Court dismissed these claims, reasoning that to decide otherwise would be to interfere with the academic decisions of the university, to encourage a pattern of educational malpractice against schools and residency programs each time a resident is sued, and to unnecessarily increase malpractice litigation if such a tort were recognized.[341] In a related case, a California Appeals court dismissed an action against a professor who was alleged to have offered an opinion regarding treatment. The court ruled that no physician-patient relationship was formed since there was no control over the actions of the actual treating doctor, and that to hold otherwise would undermine principles of academic freedom and teaching.[342]

However, in *Maxwell v. Cole*, the chairman of Obstetrics and Gynecology was successfully sued for failure to develop and enforce rules regarding qualifications and supervision of trainees. The chairman was not personally involved in the care of a woman who sustained a bladder perforation caused

---

[340] Reuter, SR. Professional Liability in Postgraduate Medical Education: Who is Liable for Resident Negligence? *J Leg Med* 1994; 15:485–531.
[341] *Swidryk v. St. Michael's Medical Center*, 493 A.2d 641 (N.J.Super. 1985).
[342] *Rainer v. Grossman*, 31 Cal.App.3d 539 (Cal.App. 2 Dist. 1973).

by resident physicians. The court disagreed with the defendant that he owed no duty because no doctor-patient relationship was formed: "*If the chief of service fails to provide medically acceptable rules and regulations which would insure appropriate supervision of ill patients, then it is reasonable to find that a breach of the standards of medical care by that individual has occurred.*"[343]

Graduate Medical Education (GME) programs that violate their own written rules naturally place themselves at risk for liability. Examples are written rules stating that catheters are to be inserted under the supervision of an attending physician,[344] or that all elective procedures are to be performed with an attending present.[345]

REFORMS TO MINIMIZE LITIGATION

Resident Work Condition — The Libby Zion Case: In 1984, an unfortunate young woman named Libby Zion presented to a New York hospital with fever and agitation, and tragically died less than 24 hours after admission with an undiagnosed illness. During her night of hospitalization, the patient was in stable condition, but early the next morning, the patient was noted to be increasingly febrile, then suffered a cardiopulmonary arrest within an hour. The intern and resident caring for Ms. Zion were questioned over issues, including delay in the patient being seen, the use of restraints, lack of supervision, the use of Meperidine in a patient who was taking an MAO-inhibitor, and failure to make a diagnosis. A malpractice suit against the residents was followed by hearings before a Manhattan Grand Jury to determine possible criminal wrongdoing. Although the residents were charged with 38 acts of gross negligence and incompetence, the Hearing Committee unanimously found that none of the charges were supported by evidence. Nonetheless, the New York State Board of Reagents voted to censure and reprimand the residents for grossly negligent care.[346]

This case alerted the nation to the issue of resident work conditions and had a great impact on graduate medical education. The Grand Jury sharply

---

[343] *Maxwell v. Cole*, 482 N.Y.S.2d 1000 (N.Y.Sup. Ct. 1984).
[344] *Siebe v. University of Cincinnati*, 766 N.E.2d 1070 (Ohio Ct.Cl. 2001).
[345] *Felice v. Valleylab, Inc.*, 520 So.2d 920 (La. Ct. App. 1988).
[346] Spritz, N. Oversight of Physician's Conduct by State Licensing Agencies: Lessons from New York's Libby Zion Case. *Ann Int Med* 1991; 115:219–22.

criticized the inadequate supervision of the housestaff during Ms. Zion's hospitalization, and urged residency program reforms. In response, the commissioner of health created the Bell Commission, which found that "*inadequate attending supervision, combined with impaired housestaff judgment due to fatigue, were contributory causes of the patient's death.*" In 1988, the New York State Health Code implemented recommendations from the Bell Commission, limiting weekly work hours to 80 hours, and consecutive hospital duty hours to 24 hours. These reforms were soon adopted nation-wide, with the intent of minimizing fatigue-related errors in healthcare.[347]

The 2003 ACGME work-hour regulations are expected to result in better-rested residents who are consequently better able to perform their duties. However, it is unclear whether they will actually decrease the amount of errors and malpractice.[348] One study of intensive care units suggested that interns made substantially more serious errors when they worked frequent shifts of 24 hours or more than when they worked shorter shifts.[349] Many observers of the transformation of work conditions contend that these regulations are actually increasing the number of medical errors because of the increased need for cross-coverage. Transferring patient information, sometimes daily, among different covering residents may be a set-up for misinformation and lapses in patient care, and accordingly, legal liability.

Substance Abuse: As many as one in ten physicians suffer from substance abuse. Residents are subjected to abnormally high levels of stress during their training, and unfortunately some resort to alcohol and drug use. The actual number is uncertain, but in one study, 12% of residents reported an increase in the use of alcohol, cocaine, or marijuana, and 7% increased their use of sedatives, opiates, or stimulants compared to the period prior to residency training.[350] In another survey, 3,000 third-year residents were found to have higher rates of alcohol and benzodiazepine use than their non-physician peers.[351] Abuse of these substances sacrifices health and functionality, and

---

[347] Brensilver, J et al. Impact of the Libby Zion Case on Graduate Medical Education in Internal Medicine. *Mt Sinai J Med* 1998; 65:296–300.
[348] Fletcher, KE et al. Systematic Review: Effects of Resident Work Hours on Patient Safety. *Ann Int Med* 2004; 141:851–7.
[349] Landrigan, CP et al. Effect of Reducing Interns' Work Hours on Serious Medical Errors in Intensive Care Units. *New Engl J Med* 2004; 351:1838–48.
[350] Koran, L and Litt, I. House Staff Well-Being. *West J Med* 1988; 148:97–101.
[351] Hughes, et al. Resident Physician Substance Abuse in the United States. *JAMA* 1991; 265:2069–73.

is potentially very dangerous in the healthcare environment. Drug abuse can increase exposure to malpractice litigation, and resources should be made available to trainees to address issues of substance abuse and dependence.

Medico-legal Curriculum: The legal aspects of medicine should be incorporated into trainee education. In 1996, the Pennsylvania Medical Society Young Physicians Section created such a curriculum by developing a handbook with contributions from clinician educators, attorneys, and risk managers. Entitled '*The Resident Physician's Practical Guide to Legal Medicine*,' it addresses medicine and the law, medical liability and malpractice, and the legal aspects of the business and administration of medicine.[352] This became the guide for teaching medico-legal issues to Pennsylvania's medical residents, and the Society's Outreach Program developed a companion seminar program to further educate its doctors.

A different approach is to offer an elective rotation at a medical liability insurance company. In one such example, emergency medicine residents were allowed to sit in on settlement discussions and reviewed thirty closed malpractice claims. During the rotation, residents were taught about claims-made versus occurrence malpractice coverage, effective documentation, medical practice issues, and system failures. The residents felt the rotation was an "*invaluable and a good learning experience*," and one resident recommended that "*All MDs should do this in their training.*"[353] While there is always a tension involved in adding to the curriculum, initiatives such as these will hopefully begin a trend for greater emphasis on medico-legal education and risk management programs for medical trainees as well as all practicing physicians.

---

[352] Kolas, C. Medicolegal Program for Resident Physicians. *Penn Med,* September 1997, pp. 28–9.
[353] Houry, D and Shockley, L. Evaluation of a Residency Program's Experience with a One-Week Emergency Medicine Rotation at a Medical Liability Insurance Company. *Acad Emerg Med* 2001; 8:765–7.

SUMMARY POINTS

MEDICAL TRAINEES

- Like all other physicians, medical trainees are susceptible to medical malpractice claims.

- Medical trainees are usually held to the standard of care expected of a duly qualified practitioner, rather than to a lower standard.

- The supervising physician and the employing hospital may be held vicariously liable for the negligent acts or omissions of trainees.

- Training programs should emphasize medico-legal education and risk management techniques to better prepare their graduates.

SECTION II

RISK MANAGEMENT

17

Medical Records and Confidentiality

Well-kept records provide the healthcare provider with more than an accounting of the patient's medical condition. Not only do they serve to document diagnosis and treatment, they also preserve discussions of risks, options and consent. A newcomer to the healthcare team will have the necessary information to assure continuity of care. Patient records also provide data for research, education, and quality assurance. Finally, records are important for business purposes, such as billing.

The medical record is of extreme importance in litigation, which typically takes place many months or years after the incident when memories regarding what was said or done may have faded. Furthermore, what was not documented can be legally construed to mean that it was not done. Whether or not the records are to be admitted as evidence is a matter for the discretion of the court, i.e., the judge. However, in the meantime, the records allow the attorneys to prepare their case. Generally speaking, entries in the medical records are admissible under the Federal Rules of Evidence as documentary evidence, so long as they are relevant and authenticated. Duplicates are admissible unless the authenticity of the original is challenged, or under the circumstances, it would be unfair to admit the duplicate in place of the original.[1]

Anything arising out of the doctor-patient encounter can constitute part of the patient's medical record. This includes items like hand-written, typed, or electronic clinical notes; notes recorded from telephone conversations; all correspondence including letters to and from other healthcare professionals, insurers, patients, family, and others; laboratory reports; radiographs and other imaging records; electrocardiograms and printouts from monitoring

---

[1] Federal Rules of Evidence 1003.

equipment; audiovisuals; and other computerized/electronic records, including e-mail messages. This last category is assuming increasing importance as the favored mode of communication, and carries with it special medico-legal risks (see Chapter 18: Cyber-Medicine).

HEALTH INSURANCE PORTABILITY AND ACCOUNTABILITY ACT OF 1996 (HIPAA)

Comprehensive federal laws that govern all aspects of a patient's medical information have recently gone into final effect. In 1996, the federal government released preliminary regulations governing the protection of patient privacy. Widely known as HIPAA, these rules have sweeping edicts that affect the handling of all medical records. The Act, a weighty document, attempts to strike a balance between the rights of patients to privacy to their medical information and the smooth and efficient delivery of healthcare. The law, which interestingly exempts those who do not utilize electronic communications, including billing, in their offices, went into effect in April 2003. HIPAA overrides state statutes that are less protective of patient access and privacy, but stricter state requirements continue to have priority.[2] States have enacted their own statutes relating to medical records, and it is therefore necessary for physicians to consult their state statutes to ensure full compliance with the law.

The intent of HIPAA is to inform consumers how their health information is being used. The emphasis is on preserving privacy of protected health information, and ensuring security of electronic transmission of such data. In practice, this means informing all patients, via postings in the office or hospital, direct mailings, brochures, etc., of the conditions under which the contents of their records will be shared with others, and the procedures in place to prevent improper disclosures.

HIPAA creates new criminal and civil penalties for improper use or disclosure of information. Fines start at $100 for each violation. Criminal penalties are severe for wrongful disclosures, with fines up to $50,000 and up to a year in prison. If the violation is committed under false pretenses, the penalty can reach $100,000 and five years in prison. Where the disclosure is coupled

---

[2] Public Law 104-191; 45 C.F.R., Part 160 and Part 164. Detailed information on HIPAA is available at the Health and Human Services website, www.hhs.gov.

with the intent to sell or to use for commercial advantage or personal gain, the punishment reaches $250,000 and up to 10 years in prison.

Some in the profession view HIPAA as excessively burdensome and bureaucratic, as it may interfere with important aspects of the traditional practice of medicine. Congress, however, passed this law in response to the perceived widespread violation of the public's privacy rights in their own medical information.

LEGIBILITY OF MEDICAL RECORDS

Bad handwriting is legendary in the healthcare profession. Medical students did not start out by having poor penmanship. Rather, as medical students graduate into practicing doctors, they quickly experience the time-pressures associated with a heavy clinical load, so charting is given short shrift. Something more pernicious has happened. The illegible handwriting is now expected, joked about and even generates perverse pride. The truth of the matter is that illegible handwriting makes a mockery of medical note-keeping, as it cannot be deciphered by others, places patients at risk when orders or prescriptions are illegible, and exposes the doctor to malpractice liability.

Studies have clearly identified poor handwriting as a source of medical errors deserving of urgent attention. Electronic medical records (EMR) will solve this problem, and will be a positive step towards greater patient safety. Of course, EMR will not do away with incorrect input or orders or other system errors, but errors from illegibility will no longer be at issue. Each year, more than 150 million phone-calls are made to doctors' officers from pharmacists who cannot make out the prescriptive order because of illegibility. Commercial firms are coming to the rescue. For example, Allscripts Healthcare Solutions, a leading commercial provider of point-of-care medication management solutions, offers an e-prescribing software called TouchScript that allows doctors to check on medication information and send their prescriptions safely and directly to the pharmacy.[3]

---

[3]Allscripts Healthcare Solutions is headquartered in Libertyville, just outside of Chicago, IL. Its website is at www.allscripts.com.

In the meantime, hospitals have resorted to temporary makeshift remedies. In some places, all notes are dictated, which is an effective but expensive and inefficient method. Other hospitals have taken the step of identifying the worst perpetrators, demanding that their notes and orders be typewritten. Some have even insisted on remedial writing classes.

A recent case serves to highlight the medico-legal risks associated with bad handwriting. The case involved a West Texas cardiologist who was ordered to pay $450,000 to the family of a cardiac patient who developed a heart attack one day after receiving a wrong medication. The patient died several days later. The pharmacist had given the patient the drug Plendil, an antihypertensive agent, instead of Isordil, an anti-anginal drug. The dose was 20 mg every six hours (the maximum dose for Plendil is 10 mg daily). The pharmacist had misread the doctor's order because of illegibility, and thought that the prescription was for Plendil instead of Isordil. Although the defense argued that the prescription error did not cause the death of the patient, the jurors concluded that the prescription error had led to a big overdose, and so had something to do with the patient's death.[4]

ACCESS TO MEDICAL RECORDS

State law requires providers to retain medical records in the original or in an unalterable computerized or miniaturized form. In Hawaii, this period of time is at least seven years from the last entry, and longer in the case of minors. After the seven-year period, the records may be destroyed, but certain basic information such as the diagnoses, operative pathology, and hospital discharge reports are held for 25 years. Many hospitals maintain records indefinitely. When patients request records, the provider should provide copies and retain the original. Stonewalling, delaying, or obstructing access to medical records are simply interpreted by plaintiff attorney as an admission of having something to hide, and hinders rather than promotes the defense against a contemplated lawsuit.

The contents of the medical record, i.e., the information contained therein, rightfully belong to the patient, although the actual hard copies are the property of the healthcare provider. Irrespective of how ownership of records is viewed, there is general agreement that the patient or his or her lawful

---

[4] *Teresa Vasquez et al. v. Ramachandra Kolluru*, Ector County (TX) District Court Case No. A-103,042 (1999).

representative usually has the right of full access to the records. This right of access is captured in Hawaii law under Hawaii Revised Statutes §622-57, which is typical of most state statutes on this matter:

> *"If a patient of a health care provider . . . requests copies of his or her medical records, the copies shall be made available to the patient unless in the opinion of the health care provider it would be detrimental to the health of the patient to obtain the records. If the health care provider is of the opinion that release of the records to the patient would be detrimental to the health of the patient, the health care provider shall advise the patient that copies of the records will be made available to the patient's attorney upon presentation of a proper authorization signed by the patient."*

HIPAA clearly establishes the rights of patients to fully access their records. Indeed, not only may they request to view and/or copy their records, they have the right to contest entries that they disagree with and to make the necessary amendments or corrections.

Occasionally, the healthcare provider may properly deny patient access to their own records. A common reason is where the information will prove injurious to the patient because of its sensitive nature. Psychiatric records, more than other types, may belong in this category. Hawaii requires handing over of the records to the patient's attorney under such circumstances. In other words, patients who want to see their records but who were denied access will succeed if they hire an attorney to get them, and the law requires the production of medical records within ten days.

RELEASE OF RECORDS

Who else is entitled to the records when there is no patient consent for release? The HIPAA rules are explicit on the scope of disclosure of protected health information. The rules limit its release without authorization by the patient or legal representative. Initially, this was to cover electronic communications, but the law was subsequently extended to cover all forms of personal medical records, including paper records and oral communications. However, all healthcare professionals involved in the treatment of the patient will have complete and total access to patient information without the need to obtain patient authorization. This applies to trainees including medical students. However, the rules restrict access by office, hospital, and other support staff on a need-to-know basis. This has been termed the 'minimum necessary standard.' Disclosures are also permissible where threats to public health or public safety are concerned.

Disclosures to other persons or entities such as insurance carriers and employers can be made only with prior patient authorization, and providers are obligated to track the movement of all personal health information. Under most state laws, medical information should generally not be released to employers, insurance carriers, other payers, researchers and others in the absence of a signed consent by the patient.

Although the patient does not own the medical records, he or she owns the information contained in them, and can therefore insist that the information be kept confidential. Medical practitioners are under no obligation, and must refuse, to produce or surrender these records to third parties, including the police, in the absence of a valid authorization — a court order, search warrant or specific law mandating release. A good practice is therefore to insist on a signed authorization before releasing privileged information.

CONFIDENTIALITY

Federal and state laws as well as all medical and nursing codes of ethics speak to preserving the confidentiality of patient information. The doctor-patient relationship is a trust relationship, which means patients can feel free to discuss personal and sometimes sensitive details of their lives and illnesses with the doctor, confident that such information will be kept confidential. Indeed, Hawaii's Rules of Evidence, which incorporates a doctor-patient privilege, immunizes the doctor from testifying about the condition of the patient unless the latter is a litigant and places the medical condition in controversy.[5]

Historically, patient confidentiality was part of an ethical code that all doctors abided in, the tradition dating back to Hippocrates:

> *"Whatever in connection with my professional practice, or not in connection with it, I see or hear, in the life of men, which ought*

---
[5] Hawaii Revised Statutes, Chapter 626, Hawaii Rules of Evidence, Rule 504.

not to be spoken of abroad, I will not divulge as reckoning that all should be kept secret."[6]

In a hospital setting, the patient's medical record is viewed by many individuals, some of whom may not even be directly involved in the patient's care. This number has increased substantially under today's managed care/peer review/utilization mandates. Even in 1982, Dr. Siegler at the University of Chicago was *"amazed to learn that at least 25 and possibly as many as 100 health professionals and administrative personnel at our university hospital had access to the patient's record and that all of them had a legitimate need, indeed a professional responsibility, to open and use that chart. Six attending physicians, 12 house officers, 20 nursing personnel, 6 respiratory therapists, 3 nutritionists, 2 clinical pharmacists, 15 students, 4 unit secretaries, 4 hospital financial officers and 4 chart reviewers."* On this basis, Dr. Siegler has termed confidentiality in medicine *"a decrepit concept."*[7]

Being a doctor does not automatically authorize one to look into every patient's chart. Doctors have no business looking into the medical records of patients who are not under their care. It is not infrequent for doctors to look at the records of a hospitalized friend or relative without being given specific authorization. This is an invasion of patient privacy.

Using the medical records for teaching purposes is permissible if the patient, recognizing that it is a teaching hospital, has given expressed or implied consent. However, such disclosures should be, but are not always, made clear to the hospitalized patient.

Confidentiality goes beyond medical records. As a whole, healthcare professionals need to be far more circumspect in observing confidentiality. Unfortunately, carelessness is not uncommon. Loose talk, gossip and disclosure of patient identity, in addition to the indiscriminate release of a patient's medical records are all examples of confidentiality breach. Healthcare professionals commonly discuss patient problems, sometimes identifying the patient by name or room and bed number, in public places within earshot of strangers. In one 1995 study that observed 259 one-way elevator trips in five

---

[6] The Hippocratic Oath can be found in Temkin, O and Temkin, CL (eds), *Ancient Medicine: Selected Papers of Ludwig Edelstein*, John Hopkins Press, Baltimore, 1967, p. 6.
[7] Siegler, M. Confidentiality in Medicine — A Decrepit Concept. *N Engl J Med* 1982; 307:1518–21.

U.S. hospitals, the authors overheard a total of 39 inappropriate comments which took place on 36 rides (13.9% of the trips).[8] Many of the comments clearly breached patient confidentiality.

LEGAL REMEDIES FOR BREACH OF CONFIDENTIALITY

The plaintiff may sue the healthcare provider based on several legal theories. These include the intentional infliction of emotional distress, medical negligence, invasion of privacy, breach of confidentiality and breach of contract. However, in most instances, the court looks at the community practice, the ethical code, and importantly, statutory provisions to determine whether there is liability.

Although both privacy and confidentiality share in common the right of the patient to control information, an important distinction is that only the one who holds information in confidence can be accused of breach of confidentiality, whereas anyone can perform an act that qualifies as a tortious invasion of privacy. The latter tort is described by Prosser, a respected legal academician, as consisting of four separate claims:

(1) Appropriation of the plaintiff's name or likeness;
(2) Unreasonable and offensive intrusion upon the seclusion of another;
(3) Public disclosure of private facts; and
(4) Publicity which places the plaintiff in a false light in the public eye.[9]

In addition to these legal remedies available to the plaintiffs, medical practitioners now face both civil and criminal sanctions if they are found in violation of HIPAA. In a civil suit for malpractice, these rules can be used to define the standard of care, and the plaintiff is likely to succeed by showing that the defendant's conduct was in violation of the HIPAA statute.

DISCLOSURE TO THIRD PARTIES

The professional duty of protecting the confidences of a patient is not an absolute one, and there are circumstances provided by law where limited

---

[8] Ubel, PA et al. Elevator Talk: Observational Study of Inappropriate Comments in a Public Space. Am J Med 1995; 99:190–4.
[9] Prosser and Keeton, Torts 851, §117 (5th Ed., 1984).

disclosure of patient secrets is not only permissible but may be obligatory. An obvious exception is a state law requiring the disclosure of communicable diseases. Other examples are the mandatory reporting of suspected child abuse and gunshot wounds.

A more difficult situation arises where there is no statute on point, but the public welfare is at stake. The well-known *Tarasoff* case established that where there is threatened harm to a named third party, the practitioner is required to reveal the information to the intended victim. This has been termed the *Tarasoff* rule, after the California case of *Tarasoff v. Regents of University of California*. In *Tarasoff*, a jilted patient confided in the University psychologist his intention to kill his ex-girlfriend. The information, though shared with campus security, was not released to the intended victim, the girlfriend, who was stabbed to death by the patient two months later. The Court found the psychologist and the University of California (under *respondeat superior*) liable, and explained that in this case the protection of public safety was more important than the sanctity of the doctor-patient confidentiality relationship:

> *"We recognize the public interest in supporting effective treatment of mental illness and in protecting the rights of patients to privacy and the consequent public importance of safeguarding the confidential character of psychotherapeutic communication. Against this interest, however, we must weigh the public interest in safety from violent assault.... In this risk-infested society, we can hardly tolerate the further exposure to danger that would result from a concealed knowledge of the therapist that his patient was lethal. If the exercise of reasonable care to protect the threatened victim requires the therapist to warn the endangered party or those who can reasonably be expected to notify him, we see no sufficient societal interest that would protect and justify concealment. The containment of such risks lies in the public interest."* [10]

Disclosure to third parties of sensitive patient information such as having a contagious disease may constitute defamation, which is defined as harming a person's reputation through communicating to others a verbal or written falsehood. However, there may be a qualified privilege to such disclosure where

---

[10] *Tarasoff v. Regents of University of California*, 551 P.2d 334 (S.Ct. Cal. 1976).

there is a public health interest. For example, in *Simonsen v. Swenson*, a physician disclosed a positive syphilis test result that turned out to be an error. In the defamation suit that followed, the court declined to impose liability on the doctor, finding that he was protected in discharging his duty to disclose.[11]

SUBPOENAS

A subpoena is nothing more than a legal request. A subpoena to produce documents including medical records is termed a subpoena *duces tecum*. A court order (sometimes incorrectly called a court-ordered subpoena), i.e., one signed by a judge, to produce clinical records must be complied with or the practitioner risks being found in contempt of court. However, not all subpoenas are the same. Those that originate from a lawyer's office, even if they appear official and are signed by the clerk of the court, do not have the same force of law as those that are court ordered. There is no legal obligation to release a patient's medical records upon being served a subpoena by the attorney that does not have the patient's signed authorization to go along with it. Indiscriminate release of records violates patient confidentiality, and a New Jersey appellate court recently ruled against a physician for disclosing medical records in response to an improper subpoena without prior patient authorization.[12] A doctor may challenge a subpoena that is unaccompanied by patient authorization to release information, but he or she should not simply ignore the subpoena. Refusal to produce may force the attorney to obtain a court order, at which time the practitioner is legally obligated to produce the records.

Some types of information are legally protected from 'discovery,' the legal process where the litigants assemble and share all the evidence prior to trial. For example, peer review deliberations are protected from discovery in a lawsuit. This rule serves to protect the confidentiality of such activities so that participants can speak freely and honestly about the case before them.

Under HIPAA, a physician is allowed to release medical records in response to a subpoena if the patient is first notified. In response to a court order in a judicial proceeding, disclosure does not require the patient's prior authorization.

---

[11] *Simonsen v. Swenson*, 177 N.W. 831 (Neb. 1920).
[12] *Crescenzo v. Crane*, 796 A.2d 283 (N. J. Super. 2002).

SUMMARY POINTS

MEDICAL RECORDS AND CONFIDENTIALITY

- Medical records serve not only to preserve medical information for diagnosis and treatment, but also to provide the most important defense tool in a malpractice suit.

- Heed the axiom: 'If it's not documented, it was not done.'

- The practitioner (or healthcare facility) owns the physical records; the patient owns the information contained therein.

- Retention and release of records are governed by state and federal statutes, including the strict HIPAA rules.

- Illegible handwriting is a curse upon the profession. Illegible orders and prescriptions are a curse upon the patient.

18

Cyber-Medicine

Virtually everyone these days is caught up in the stampede over information technology (IT). IT, especially in the form of the Internet, is supposed to transform the world as we know it with untold benefits for mankind. In no other field of human endeavor is this promise more enticing than in healthcare. We are told that IT will directly benefit our patients. This presumably means patients will live longer and healthier, and that healthcare will be delivered more efficiently, affordably and competently. Patient satisfaction should soar.

Whether IT will improve the doctor-patient relationship is doubtful. What is not in doubt, however, is the medico-legal liability that attaches to medical activities conducted over the Internet, i.e., cyber-medicine. This area of the law is new and litigation is certain to proliferate in the years ahead. This chapter will offer a broad overview of the legal principles of negligence and privacy as they impact medical practice in the era of telemedicine, electronic communications, and Website sales, advice, prescriptions, and consultations.

MEDICO-LEGAL RISKS

With respect to IT, medico-legal risks fall into three main areas: Breach of privacy and confidentiality, medical negligence, and informed consent. There may also be issues of products liability, cross-border jurisdictional litigation, and other liabilities.

Privacy & Confidentiality: The awesome potential of the Internet to offer large-scale shared electronic databases properly raises ethical and legal issues of confidentiality and rights of access. The whole world can now literally peek at the patient's records, unless they are secured. Projects such as the Boston Electronic Medical Record Collaborative are out there to develop reliable protocols to electronically identify patients and providers, secure permission for release of records, and track information that is transmitted.[13]

Then there is the issue of the confidentiality of e-mail messages. These messages are not as private as one would like to believe unless they are encrypted. Doctors who use emails to communicate with or about their patients are obligated to ensure the privacy of these communications.

The HIPAA rules (discussed in Chapter 17: Medical Records and Confidentiality) impose both civil and criminal liabilities on the healthcare provider who is found in violation of standards regarding privacy and confidentiality — on-line and off-line.

Medical Negligence: Claims of medical malpractice will be alleged for acts in cyberspace. The plaintiff will have to prove that there was a duty of due care that was breached. Substandard conduct may include, among other things, wrong advice, untimely diagnosis and referrals, treatment errors including prescription errors, and failure to return calls or respond to electronic messages, i.e., e-mail. Disclaimers are regularly posted by the on-line doctor-advisor-consultant to try to avoid liability. Whether this will withstand legal scrutiny is doubtful, as courts may well find that a doctor-patient relationship had indeed been formed in cyberspace. Operators of for-profit medical Websites may be particularly vulnerable for such claims of negligence.

Informed Consent: The most likely 'medical treatment' on the Internet involves getting medical advice or prescriptions, so getting consent may not appear to be that difficult. However, one has to be confident that the patient understands the issues and what is being proffered and that consent is being freely given. Doctors who wish to conduct business on the Internet should consult legal counsel before opening their Websites. Claims asserting lack of informed consent are a part of virtually every malpractice lawsuit and 'cyber-consent' may yet emerge as a minefield.

eRISK

The term 'eRisk' has been coined to describe potential malpractice risks that a healthcare provider faces when conducting patient communication,

---

[13] Rind, DM et al. Maintaining the Confidentiality of Medical Records Shared over the Internet and the World Wide Web. *Ann Int Med* 1997; 127:138–41.

education, and care over the Internet. Medem, Inc., a national coalition of medical societies covering over two-thirds of physicians, has recently published an excellent overview of the subject in a document entitled *'eRisk for Providers: Understanding and Mitigating Provider Risk Associated with Online Patient Interaction.'*[14] The document is the product of Medem's eRisk Working Group for Healthcare, which received input from malpractice carriers, medical societies, and legal counsel. The document lists the following four areas of Internet risk to providers:

1. Liability for online malpractice.
2. Liability associated with the expansion of the physician's office and the patient-physician relationship onto the Internet.
3. Liability associated with the inadvertent creation of a physician-patient relationship or other licensed provider-patient relationship.
4. Liability for failures of process and/or supporting technology resulting in misuse, inappropriate disclosure, loss, or inappropriate access to patient information (including issues of HIPAA and other regulatory compliance).

Whether a doctor-patient relationship can be established on-line remains an unanswered legal question. However, it is likely that such a relationship will be found in some circumstances. Case law governing telephone contacts examines issues such as whether the doctor and patient saw each other, whether there was knowledge of the patient's name, whether there was ever a physical exam, whether the doctor saw the records, whether there was payment, and whether the doctor accepted a request for an appointment. The more 'yes' answers to these questions, the more likely a relationship was formed, giving rise to a duty of due care. If on-line interactions are held to offer more information-rich interaction than the telephone, as is likely, doctors may be deemed to have constructively formed a relationship with his or her cyber-patient, even if there had been no physical contact or real-world interaction.

Some providers have chosen to produce, publish and maintain their own Websites. Patients go to the Internet for clinical information and some are

---

[14] Portions used with permission of Medem, Inc., and the Risk Working Group for Healthcare. Address of Medem, Inc. is 333 Bush Street, 24th Floor, San Francisco, CA 94104. Website is www.medem.com/erisk. Its email address is info@medem.com.

asking their physicians for guidance. Doctors are either providing this information on their own Websites or directing their patients elsewhere for it. In both situations, issues arise regarding the content source and responsibility for updates. There are also potential risks of conflict of interests associated with delivery of advertising/sponsorship along with the clinical information. The eRisk Working Group feels that if a physician puts clinical information on his or her Website, then there should be an affirmative obligation to ensure currency and high quality.

Physicians and patients are increasingly using e-mail to exchange confidential healthcare information. The difficulty is that adoption of new technologies is outpacing the development of standards that dictate proper use. The eRisk Group supports the concept of electronic 'P2P' (provider to patient) communications, but cautions that the provider is at increased risk if he or she uses an unauthenticated, non-encrypted, non-secure communication network. Most readily available e-mail providers do not currently meet this standard.

There is also the issue of using Internet pharmacies for dispensing medications. Online pharmacies have proliferated, and most of them are legitimate and provide an excellent service to patients. Obvious benefits include easier access, efficiency, lower cost, and anonymity. However, some cyberpharmacies may employ 'cyberdocs' in order to dispense without a valid doctor's order. Others may be involved in illegal importation of prescription drugs. Federal and state agencies have enacted regulations to rein in questionable, unsafe or illegal practices. Clinicians who refer their patients to on-line pharmacies, or agree to do so upon their request must be aware of these risks.[15] The National Association of Boards of Pharmacy provides a Verified Internet Pharmacy Practice Sites (VIPPS) program that certifies pharmacies as being in compliance with standards that include critical on-site inspection and review. Such Internet pharmacies will feature a seal of approval on their home page.

---

[15] Fong, CH et al. Controversies and Legal Issues of Prescribing and Dispensing Medications Using the Internet. *Mayo Clin Proc* 2004; 79:188–94.

SUMMARY POINTS

CYBER-MEDICINE

- Medical activities over the Internet carry significant medico-legal risks.

- Liability can result from invasion of privacy, breach of confidentiality, negligence, and lack of informed consent.

- Healthcare providers should be aware that a doctor-patient relationship may be created in cyberspace, thereby imposing a duty of due care.

- Reasonable precautions to safeguard confidentiality must be provided for e-mail and other Internet communications.

- Clinicians should be aware of the risks of referring patients to on-line pharmacies.

19

What Do Malpractice Lawsuits Look Like

Healthcare professionals are forever telling and re-telling horror stories about malpractice cases and their accompanying outlandish awards. However, usually not all the facts of the cases are known or told and the damages are frequently exaggerated or misrepresented. Yet some of these stories, shocking though they may sound, do represent actual true-life cases. This chapter looks at the frequency and severity of claims and outlines the nature and outcome of the most dramatic cases. In real life, most malpractice cases are rather straightforward and damages rarely reach or exceed the $1 million mark. Negotiated settlements are naturally lower than those awarded at trial (see Table 9).

FREQUENCY AND TYPES OF CLAIMS

According to the National Association of Insurance Commissioners, the total number of new medical malpractice claims reached 90,212 in 1995; this figure declined slightly to 86,480 in 2000.[16] Before 1960, one in seven doctors had been sued during their entire careers. However, it has been stated, not without challenge, that currently one in seven doctors (one in six according to the AMA) is sued every year.[17] The risk exposure of the various medical and surgical specialties to malpractice lawsuits has changed little over time, with the notable exception of anesthesiology, where the number of claims has nose-dived. This decline is attributed to the adoption of patient-centered

---
[16] Cited in Malpractice Statistics, available at www.medical-malpractice-lawyers-attorneys.com/medical_malpractice_statistics.htm. Accessed March 9, 2004.
[17] Kereiakes, DJ and Willerson, JT. Health Care on Trial: America's Medical Malpractice Crisis. *Circ* 2004; 109:2939–41.

safety measures to prevent operative anesthetic harm. Such measures include pulse oximetry, capnography, and other perioperative monitoring devices and standards.[18]

Comprehensive data from 25 medical malpractice insurers were collected in 1984 by the U.S. General Accounting Office (GAO). Although over 20-years old, these data regarding malpractice claims are not very different from today's experience except that allegations of lack of informed consent have now become an important component of many malpractice lawsuits.

The GAO study showed the following distribution and types of claims:

- **Surgery-related (25.4%):** Leading events included post-op complications, inadvertent acts, inappropriate procedures, post-op death, and unnecessary surgery.
- **Diagnosis-related (23.6%):** These included failure to diagnose, misdiagnosis, failure to obtain consent, unnecessary diagnostic tests, and improper performance of testing. Some of the leading 'failure to diagnose' claims involved cancer, circulatory problems such as myocardial infarct, fractures and dislocations, and infections.
- **Treatment-related (19.9%):** Examples included improper treatment during child delivery, insufficient therapy, improper treatment of infection, and drug side effects.
- **Medication-related (7.8%):** Cases involved the administration of the wrong drug, at the wrong dosage, to the wrong patient, or via the wrong route, and overlooked drug allergies. There was a sharp increase in this area of litigation.

SEVERITY OF CLAIMS

Depending on the source, one can gather rather different data regarding the quantum of malpractice damages. It is important to note the specialty being referred to, e.g., internists versus surgeons, and to distinguish between settlement figures and jury awards.

---

[18]Arnold, TV. The Anesthesiology Claim Experience: A Success Story. *N J Med* 2000 (December), pp. 35–38.

The listing below (Table 7) shows the average damages for various injuries incurred in the 1980s.[19] At the time, these amounts were considered outrageously high, but they pale in comparison with more recent figures.

Table 7: Malpractice Damages in the 1980s

Neurological deficit and other birth-related injuries	$276,170
Cancer	$172,408
Myocardial infarct, cardiac arrest, and heart disease	$154,099
Fractures and spinal problems	$110,945
Medication-related claims	$94,838

By 1994, the median jury award (not settlement amounts) in medical malpractice suits had risen to $200,000. This compares with $30,000 for vehicular-accident suits. The highest median award was paid by anesthesiologists at $868,804. Although the frequency of claims against anesthesiologists has decreased, their severity remains high because of the catastrophic nature of anesthetic mishaps. Between 1990 and 1993, there were 397 awards that exceeded a million dollars each.[20]

Even these figures have become obsolete. One in four jury verdicts in 1999 reached $1 million. In Pennsylvania, for example, there were 19 such awards in 1998, but in 1999, there were 33. Even in California, where tort-reform is in place, the number of claims with awards or settlements of over $1 million has soared. These claims were in the single digit range in the 70s, in the tens in the 80s, and have steadily increased during the last two decades, reaching 40 in 2002 and 48 in 2003. The median award rose to $2.8 million in 2003, whereas it was $1.57 million in 2001. According to the Medical Underwriters of California, "*The 14 highest value cases in 2003 each involved present-value awards of more than $3 million, and totaled $86.1 million.*"[21]

Nationwide data does not always support such upward trends. According to a *Wall Street Journal* article, the latest figures from Jury Verdict Research showed stability from 2000–2002. In 2002, the median award was

---

[19] Physician Insurers Association of America, Data Sharing Reports, 1985–1991, Pennington, N.J., June 1991.
[20] *Smart Money*, August 1995, p. 76.
[21] Medical Insurance Exchange of California's website, MIEC.com. Accessed June 15, 2004.

$1.01 million, only up slightly from $1 million in 2000 and 2001. The amount in 1996, however, was less than half at $473,000. In the past 4 years, approximately half of all awarded damages were in excess of $1 million, and this percentage has not changed. Interestingly, for all personal injury liabilities, median awards actually fell to $30,000 in 2002, about 30% lower than the $43,000 figure in 2001.[22] In its latest release, Jury Verdict Research reported that the median malpractice award in 2003 had risen to $1.2 million.[23]

Data on 1999 damages for site-specific, personal injuries are shown in Table 8. These figures represent median damages derived from 176,000 jury awards and out-of-court settlements.[24]

Table 8: Site-Specific Damages (1999)

Severe Brain Damage	$4,500,000
Spinal Nerve Injury	$834,000
Intestinal Injury	$612,000
Eye Injury	$500,000
Leg Injury	$500,000
Genital Injury	$250,000

Table 9: National Median Jury Awards and Settlements

	Jury Awards	Settlements
1993	$500,000	$400,000
1994	$375,000	$287,000
1995	$500,000	$340,000
1996	$454,565	$355,000
1997	$500,725	$400,000
1998	$750,000	$500,000
1999	$800,000	$650,000

---

[22] Malpractice Awards Remain Flat. *The Wall Street Journal*, April 1, 2004, p. D-4.
[23] See JVR press releases at www. juryverdictresearch.com. Accessed on November 27, 2005.
[24] Jury Verdict Research, "2000 Current Award Trends in Personal Injury."

According to *The Wall Street Journal's* compilation of the latest statistics from various governmental sources, the amount of healthcare spending accounted for by malpractice costs is 2%, and the median payment for damages on behalf of physicians was $150,000.[25]

HORROR STORIES

To convey a sense of what serious malpractice cases might look like, I have listed below consecutive lead cases (2000–2002) published in *Legal Review & Commentary*, a monthly supplement to the trade newsletter, *Healthcare Risk Management*.[26] These cases reflect the most dramatic worst case scenarios rather than typical claims. I have retained the headlines, followed by a brief summary of the salient features of each case. The reader is encouraged to review these case reports in their entirety, as published in *Healthcare Risk Management*.

Remember that the actual damages paid out by the defendant in any given case are typically smaller than verdict awards. For example, *The Wall Street Journal* recently recounted the case of a $269 million Texas jury award with 25% or $67 million apportioned to three doctors who had allegedly administered ten times the usual dose of the sedative Propofol. The overdose caused rhabdomyolysis, renal failure and death in a 15-year-old girl with cerebral palsy. However, the parents received only $3 million from the three physician defendants per settlement that was reached just before the jury returned its verdict.[27]

- **Patient burns to death: Jury awards $1 million verdict.** A 56-year-old woman with a history of bipolar disorder was allowed to smoke, and she used her cigarette lighter to open a package of crackers, setting fire to the plastic wrap and her nightgown. She died 10 days later from the burns.

- **Patient's daughter alleges inadequate medical care: Florida nursing home agrees to $1.1 million settlement.** The daughter of a nursing home patient,

---

[25] In Malpractice Trials, Juries Rarely Have the Last Word. *The Wall Street Journal*, November 30, 2004, p. A-1.
[26] *Healthcare Risk Management*, 3525 Piedmont Road, Building Six, Suite 400, Atlanta, Georgia 30305. www.hrmnewsletter.com. Tel. (800) 688-2421.
[27] In Malpractice Trials, Juries Rarely Have the Last Word. *The Wall Street Journal*, November 30, 2004, p. A-1.

an 83-year-old with Alzheimer's disease, successfully sued the facility alleging that the nursing home personnel failed to adequately prevent and treat the patient's infected skin ulcers and dehydration.

- **Failure to communicate need for cancer tests: $850,000.** The plaintiff patient and her husband received a jury verdict of $850,000 against a physician who failed to communicate the need for a colonoscopy after an endoscopy showed no ulcer despite bleeding, and did not contact the patient when the preoperative blood work indicated anemia. Colon cancer was diagnosed a year later and had progressed significantly.

- **Cover-up of failed surgical repair: $16.8 million verdict.** During a thoracotomy for the removal of a lung lobe, the surgeon cut a hole in the plaintiff's diaphragm. The surgeons repaired the hole but did not report its existence. The repair did not hold, and the plaintiff required five months of hospitalization to recover from the incident. A $16.8 million verdict was returned against the two surgeons and their employers as well as the hospital.

- **Patient loses lip to frostbite: $2.55 million settlement.** A $2.55 million settlement was reached with the manufacturers of the liquid oxygen delivery base unit and flow meter, and the supplier of the liquid oxygen that caused severe frostbite to a plaintiff's face and mouth. The flow meter was not designed to read the actual flow of oxygen from the base unit, which allowed the flow to be higher than the level the machine indicated.

- **Woman denied admission to psychiatric unit, commits suicide the next morning: $6.5 million verdict awarded.** A 43-year-old woman presented herself at a hospital emergency department asking to be admitted to the psychiatric unit. The admission was denied and she committed suicide the next morning. An Illinois jury returned a $6.5 million net verdict against the hospital.

- **Chain of events leads to brain damage: $4.95 million.** A 37-year-old man suffered irreversible brain damage after a series of mishaps at a Colorado hospital. The anesthesiologist had inadvertently administered phenol instead of guanethidine, and failed to monitor the patient during various procedures in the hospital.

- **Jury finds hospital acted with malice in granting physician privileges: $40.6 million in damages set.** The jury found that Columbia/HCA Health

Care Corp. acted with actual malice in granting surgical privileges to a doctor who had been sued 11 times before the case at hand (herniated disk repair in a 40-year-old dock worker). Punitive damages were set at $12 million. In addition, the jury held that the physicians and hospital had been negligent and committed medical malpractice. Damages were set at $28.6 million.

- **Physicians ignore man's allergy, resulting in death: Michigan bench trial ends with a $150,000 verdict.** A 66-year-old Michigan man died following hospitalization and treatment with Bactrim (a sulfa-containing drug) to which he was allergic. The patient's allergy was known, and the use of Cipro would have been appropriate, but at the Veterans Affairs (VA) hospital, use of this drug required prior approval (cost per day of Bactrim was $4, for Cipro, $7).

- **Nursing home resident with Alzheimer's disease falls from window, fractures hip: $550,000 verdict.** Because of his known tendency to wander away, a 79-year-old man with Alzheimer's disease was transferred to a new nursing home purported to feature a secured Alzheimer's unit. In fact, this new nursing home had difficulties keeping its residents in house. While attempting to wander away from the facility, the patient fell from an unsecured window (facility was not air-conditioned), and broke his hip. He later died from complicating pneumonia. A Florida jury returned a $550,000 verdict against the nursing home and its parent company, which subsequently filed for bankruptcy.

- **Patient develops AIDS phobia: $25,000 settlement.** A hospitalized woman, stepping on a glass capillary tube, broke it. She brought suit against the hospital, alleging she had developed AIDS phobia. Although there was no evidence that the glass tube contained any blood, much less that it was contaminated with HIV, the hospital nonetheless settled the case for $25,000. The court threw out the patient's suit against the hospital's cleaning service contractor for insufficient evidence.

- **Equipment failure ends in brain injury: $500,000 settlement.** A 53-year-old man underwent hip replacement surgery at a VA teaching hospital. A valve on the anesthesia circuit was stuck in the closed position, and the anesthesiology resident was unable to ventilate the patient, who then suffered ischemic brain damage. The medical malpractice claim was settled for $500,000.

- **IV inserted to nerve results in radial nerve injury: $155,000 arbitration award.** The patient was a 29-year-old divorcee mail carrier and mother of two minor children. During the pre-op care for nose reconstruction surgery, an intravenous line was placed in her wrist, hitting her radial nerve. As a result, she developed reflex sympathetic dystrophy. The plaintiff alleged that this was a classic *res ipsa loquitur* case.

- **Untimely treatment of pre-eclampsia leads to death: $1.1 million verdict.** An extremely obese (350 lbs) woman, aged 26, developed hypertension in the 38th week of pregnancy. Despite hypertension, edema and proteinuria, she was not advised of the risk associated with her condition. She was neither monitored nor treated as a high-risk case. Following C-section, she developed pulmonary emboli and ARDS, dying 11 days after she delivered a health baby girl. Both the attending physician and the hospital were found liable (75% and 25%, respectively). Of the hospital's share, the jury found the nurses 10% responsible under the doctrine of vicarious liability (*respondeat superior*) and the hospital itself 15% under the theory of corporate negligence.

- **Delay in hyperbaric chamber treatment leads to paralysis: $31 million verdict.** A young diver developed the 'bends' and suffered severe injuries, including paralysis. He claimed that a series of mishaps significantly delayed his obtaining treatment for the condition. The community hospital personnel delayed the transportation, administered the wrong intravenous fluid (5% glucose instead of normal saline), and delayed hyperbaric chamber treatment. It seemed that the receiving hospital had not been contacted to prepare the chamber and assemble the requisite medical team. The defendant countered unsuccessfully with an assumption of risk defense. The jury awarded $31 million in damages, and the judge subsequently reduced the jury award to $22 million.

- **Patient falls in recovery room, lands on just-operated knee: $175,000 verdict.** A jury returned a verdict of negligence against the hospital where a 51-year-old patient fell when she tried to stand up a few hours following minor arthroscopic knee surgery. She required subsequent corrective surgery and alleged that the recovery room staff failed to properly assess her level of recovery and allowed her to stand up too soon.

- **Failure to diagnose osteomyelitis, thoracic para-spinal abscess: $800,000 Arizona verdict.** The diagnoses were missed by a number of doctors whom the plaintiff had consulted during the course of several weeks for severe back pain. Residual neurologic damage resulted (paraparesis, loss of bladder and bowel control, and sexual dysfunction). The doctors settled, but suit was continued against the hospital where the plaintiff first sought care. The jury awarded damages of $800,000.

• **Male operating room technician inappropriately touched by male surgeon: $60,000 Texas verdict.** A plastic surgeon kicked the buttocks of a young male operating room technician and then grabbed and squeezed his right breast saying: "My wife just divorced me, but I think that you will do just fine." As a result of the incident, the technician became depressed and feared working with the plastic surgeon again. A jury awarded the plaintiff $20,000 for mental anguish and $40,000 in punitive damages (this is an assault and battery action rather than malpractice).

• **Fire in a patient's throat during surgery leads to death: $2.1 million settlement.** An 83-year-old retired farmer with a cancerous vocal cord underwent laser surgery as an outpatient. The medical team used a Z-Xomed laser-resistant tube, but failed to follow the manufacturer's instructions regarding placing wet pledgets around the cuff. Moreover, the anesthesiologist delivered an excessive concentration of oxygen. As a result, the procedure caused a fire in the patient's throat that subsequently caused his death. The doctors settled the case for $2.1 million prior to trial.

• **Improper treatment of broken arm results in brain damage: $6.5 million settlement in Washington.** A 9-year-old girl fractured her forearm at school. A puncture wound that was not timely and aggressively treated complicated her condition. She developed a severe infection, hyponatremia, grand-mal seizures, and permanent brain damage. The various defendant doctors (emergency department doctor, orthopedic surgeon) and hospital each claimed that others were at fault, but eventually settled for a combined amount of $6.5 million.

• **Failure to monitor oxygen levels leads to brain damage: $2.7 million settlement in Kansas.** Nurses at an understaffed Kansas hospital failed to carry out the doctor's orders to carefully monitor a 63-year-old woman who was admitted with pneumonia and respiratory distress. The woman suffered hypoxic brain damage. The hospital contended that it was the victim of a nationwide nursing shortage that affected all healthcare facilities. When the plaintiff threatened to sue for punitive damages, the hospital settled the case for $2.7 million. It attempted to link the settlement to an agreement to remain silent, but the plaintiff refused to go along. The story made the front page of the local newspaper, and also appeared on television.

- **Complications with twins lead to $4.9 million Ohio settlement.** A small town family practitioner was sued for negligence following the delivery of identical twins, one of whom suffered severe mental retardation because of an umbilical cord prolapse. The mother had an arrest of labor, and the family practitioner had failed to call in an obstetrician for assistance. The practitioner was also slow to call in the surgical team to perform the emergency C-section. The anesthesiologist was also sued because she took 20 minutes to respond to the call although she lived only 2.3 miles away. Combined settlement of all parties, including the hospital: $4.9 million.

- **Failure to diagnose newborn's heart-valve defect: $500,000 judgment in Indiana.** A 16-year-old underwent an uneventful labor and delivery, but her child died a few days after birth from complications of an undiagnosed congenital hypoplastic left heart valve. The plaintiff alleged that she had noted bluish discoloration of the infant's limbs, but both the nurses and medical staff ignored her complaints. A jury found the hospital and its nursing staff, but not the doctors, to be at fault, and awarded the plaintiff $500,000 in damages.

- **Anesthesia pump malfunction injures patient: $350,000 verdict plus confidential settlement.** A 34-year-old licensed practical nurse underwent minor shoulder surgery and was placed on a Baxter PCA II anesthesia pump for self-administered analgesia. Because the device did not incorporate an anti-siphon valve, the patient received an overdose of Demerol over an 82-minute period. She alleged that her injuries included organic brain damage, but the defendants countered that her problem was depression. The jury returned a verdict in favor of the plaintiff, and awarded total damages of $350,000.

The latest 'horror story' is liability for failure to provide adequate pain relief. In the California case of *Bergman v. Eden*, the treating internist was found reckless for not adequately relieving pain in an 85-year-old terminally ill cancer patient. The jurors awarded the family $1.5 million in damages. The doctor's act was equated with elder abuse.[28]

---

[28] *Bergman v. Eden Medical Center*, No. H205732-1 (Sup. Ct. Alameda Co., Cal. 2001).

SUMMARY POINTS

WHAT DO MALPRACTICE LAWSUITS LOOK LIKE?

- Malpractice claims typically result from the failure to diagnose, treat, timely refer, and/or to obtain informed consent.

- Surgical complications and medication side effects are also common reasons for a malpractice claim.

- There are a plethora of horror stories where malpractice damages reached or exceeded the million-dollar mark, although the majority of cases are settled for less.

20

How to Avoid Becoming a Horror Story

This chapter will review the steps that doctors can take to minimize the risk of being sued for malpractice. Practicing one's profession with the highest competence and compassion is by far the best shield against adverse results. This means a serious acceptance of continuing medical education, an unrushed examination of the patient, obtaining informed consent, ordering appropriate tests, instituting the best evidence-based therapy, and making a timely referral to a specialist. The careful doctor avoids sloppy habits, does not cut corners, and does not work with doctors who are incompetent, else they risk being roped in as a co-defendant.

The best efforts, however, may still leave the practitioner at risk for a malpractice claim, or even losing an otherwise non-meritorious lawsuit. To minimize these eventualities, risk managers time and again stress the overriding importance of two aspects of defensive practice: Communication and Documentation.

COMMUNICATION

It has long been recognized that the quality of medical care poorly correlates with malpractice lawsuits. Instead, what prompts a claim are poor quality of communication and the perception that the physician is at fault for the bad result. The combination of bad outcome and patient dissatisfaction is a recipe for litigation.

In a study of obstetricians/gynecologists, it was found that the quality of treatment as judged by peer review was not different in frequently sued versus never sued doctors.[29] In another study on the relationship between malpractice and patient satisfaction, patients of doctors with prior malpractice claims reported feeling rushed, feeling ignored, receiving inadequate explanations or advice, and spending less time during routine visits when compared to patients of doctors without prior claims.[30] Beckman et al. have documented that communication problems exist in over 70% of malpractice cases, with these problems centering around four themes:

(1) Deserting the patient;
(2) Devaluing patient/family views;
(3) Delivering information poorly; and
(4) Failing to understand the patient/family perspective.[31]

In a telling study by Lester and Smith, the authors asked 160 adults to view a videotape of a clinical encounter that resulted in complications. In one scenario, the doctor used positive communication behaviors, and in another, negative communication behaviors.[32] They were then asked whether they would be inclined to sue the doctor.

Positive communication behaviors:

Eye contact
Friendly tone of voice
Presentation of and request for information
Smiling
Appropriate physical touch such as shaking hands
Self-disclosure
Acknowledgement of verbalizations
Reflection of affect
Appropriate praise
Relatively long period of contact (5 minutes 25 seconds)

---

[29] Entman, SS et al. The Relationship Between Malpractice Claims History and Subsequent Obstetric Care. JAMA 1994; 272:1588–91.
[30] Hickson, GB et al. Obstetricians' Prior Malpractice Experience and Patients' Satisfaction with Care. JAMA 1994; 272:1583–7.
[31] Beckma, HB et al. The Doctor-Patient Relationship and Malpractice. Arch Int Med 1994; 154:1365–70.
[32] Lester, GW and Smith, SG. Listening and Talking to Patients: A Remedy for Malpractice Suits? West J Med 1993; 158:268–72.

Negative communication behaviors:

No eye contact
Harsh and clipped tone of voice
Criticism
Minimal presentation of and requests for information
No smiling expressions
No friendly physical contact
No acknowledgements of verbalizations
No reflection of affect
No praise
Relatively short period of contact (2 minutes 2 seconds)

The videotape viewers, as a group, expressed the belief that negative communication behaviors by the physician increased litigious intentions. An increased perception of physician fault for the bad result, as well as uncertainty as to the reason for the bad outcome, also raised litigious feelings. These results prompted the authors to state:

> "...positive communications would result in less litigiousness because the physician is viewed as having cared about the patient and thus having acted in good faith. In the world of relating, good faith counts for a lot: one's reading of good and bad faith tends to define who is a malicious villain and who is a fallible human being. On the other hand, negative behaviors tend to communicate lack of concern and even antagonism and may be seen by patient as a violation of the unwritten but inherent 'caring' nature of the physician-patient relationship. Long before there is any medical outcome to be concerned about, the patient may believe that the physician has already done something 'wrong' simply by relating in what is perceived to be an uncaring manner. This may set the stage for later retaliation if something does go wrong."

The authors' advice: "*To lower litigation risk by using extra medical procedures and tests, consultation, and extensive documentation, often known as 'defensive medicine,' may miss the point. Defensive medicine is not so much a tool to prevent lawsuits as it is to win them if they do occur. But if the intention is to prevent a lawsuit in the first place, forging a physician-patient bond that can effectively resist the pressure of our litigation-crazed and socially antagonistic society seems indispensable."*

Good advice, indeed. Every effort should be made to communicate effectively, with empathy and tact. And communicating well begins with active listening.

Talking with Patients: Listening actively to patients is a basic function of a good doctor. This means more than attentive listening. The doctor should show his or her understanding of what had just been said by the patient. This reassures the patient that the doctor has heard and understood the questions or concerns. Body language is important too. Experts tell us that 55% of information conveyed in a normal conversation occurs via non-verbal means. The percentage may be even higher in a medical encounter. And being able to explain the condition, procedure, or medication in simple lay terms is critically important. Patients want their doctors to listen to them and to explain their conditions and treatment plans in simple, understandable language. During the course of a good doctor-patient relationship, the physician can effectively reduce the odds of being sued by educating the patient regarding the scope and limitations of medical care, and the patient's own responsibilities in complying with medical advice, medication, and follow-up.

The physician should give patients ample opportunity to tell their story and to ask questions. In a well-publicized study, only 23% of patients were able to complete their opening statement before the doctor interrupted, which occurred, on the average, 18 seconds after the patient began to speak![33]

The doctrine of informed consent is largely based on proper and timely communication between provider and patient. Specially prepared professional literature often helps to answer the patient's questions. Videotapes and DVDs can help to inform, and are particularly effective in explaining surgical procedures, e.g., cholecystectomy. Nothing, however, can replace the doctor-patient discussion, which should always supplement a brochure or video presentation. Such communication must be carried out at a leisurely pace, way before an elective intervention, and certainly not at the last minute along the waiting corridors of the operating suite. Avoid telling patients that

---

[33] Beckman, HB. The Effect of Physician Behavior on the Collection of Data. *Ann Int Med* 1984; 101:692–6.

the procedure is very simple, or that there will be no problems. And, of course, never guarantee results.

After obtaining the permission of the patient to avoid violating confidentiality and HIPAA rules, engage family members in the discussion if at all possible. Involving family members facilitates achieving a good clinical outcome. One ought to remember that it is the family who sues if the patient dies or suffers irreversible cognitive injuries.

Do not hesitate to call the patient or family members at home in order to remind, reassure, or clarify. This is especially important if the treatment or test-procedure had lasted longer than usual, was traumatic, complicated, or may result in post-treatment complications. For anything more than the routine, the healthcare professional or assistant should be available after hours to assist with patient concerns. Answer all patient phone calls in a timely fashion. It is usually best to make the call yourself rather than relegate it to an assistant. No patient will be unappreciative of a doctor who has taken the time to personally return his or her phone call. It means the doctor cares. And appreciative patients usually do not sue.

Sometimes physicians unintentionally use harsh or negative-sounding words. These may upset patients more than they realize. In response to this perceived need to improve interactive skills in the workplace, various groups have organized and offered communication classes and workshops. One example is *Tongue Fu!*, whose philosophy is that every time one communicates, one either sets up rapport or resentment.[34] Therefore, it is best to avoid trigger words that cause conflict and replace them with words that create cooperation.

Telephone Communication: The four basic rules are: (1) Listen and instruct carefully; (2) Insist on seeing the patient or have the patient go to the emergency department if there is any doubt; (3) Ask the patient (or pharmacist) to repeat your instructions or orders to minimize miscommunication; and (4) Document everything in writing. Risk managers warn in particular of calls concerning abdominal or chest pain, high fever, seizures, bleeding, head injury, dyspnea, tight orthopedic casts, visual complaints, and onset of labor.[35]

---

[34] Horn, S. *Tongue Fu!* St. Martin's Press, New York, 1997.
[35] Gorney, M and Bristow, J. Telephone Communication for Physicians. The Doctors Company website www.thedoctors.com. Accessed 8/22/04.

And practice consultants recommend getting rid of "confusing menus, canned ads, irksome music, assurances that 'your call is important,' and waits that border on eternity."[36] Finally, don't forget that a doctor-patient relationship may be formed as a result of a phone conversation, with an attendant legal duty of care expected of the doctor (see Chapter 2: Duty).

Tips to Defer, Deflect, and Defuse Verbal Conflict: Words can hurt, and they can heal. Too often doctors use them carelessly. Below are excerpts from a recent communication newsletter for the staff at St. Francis Medical Center in Honolulu.[37]

Use **and** (helpful) instead of **but** (hurtful):

> **But** cancels out what's been said and anchors the discussion in an argument. "I know we said it would only take 15 minutes, **but** our computer is down," or "I realize how important this is to you, **but** we tried it before..."
>
> Connect rather than cancel by using the word, **and**. "I know we said it would only take 15 minutes, **and** our computer is down," or "I realize how important this is to you, **and** we tried it before..."

Use **next time** (future) instead of **should** (past):

> No one can undo the past. If someone does something wrong and we tell them what they **should** have done, they will resent us even if what we're saying is right. We are critics when we say, "you **should** have" and become coaches instead by saying "**next time**."
>
> Focus on the future rather than punish the past. Help people learn rather than lose face. Instead of: "You **should** have made an appointment," say: "**Next time**, please call for any appointment before you come." Instead of: "You **should** have called if you were going to be late," say: "**Next time**, call me if you'll be late."

---

[36]Lippman, H. Do Patients Love You, but Hate Your Phones? *Medical Economics*, October 2000, pp. 9–12.
[37]I am grateful to Ms Gwen Clarke-Fujie, the author, for permission to reproduce the material, some of which was adapted from a seminar given by S. Horn (see footnote 34).

Use **if you would/if you could** (request) instead of **you'll have to/you need to** (order/command):

> No one likes to be ordered around. "**You have to** call your insurance company and **you need to** get their approval."
>
> Rephrase orders into requests or recommendations: "**Would you** call your insurance company, and **could you** get their approval?" People often prefer fulfilling a request or recommendation, rather than just following our orders. "Mrs. Brokenankle, when you do your exercises, **would you** do it two minutes longer each time and **could you** do a good stretch before each session?" Said in a kind, patient and gentle way, such words could create better results for everyone.

Use **Sure, as soon as** (positive) instead of **No, you can't because** (negative):

> "**No, you can't because**" is like a door slamming in someone's face. "**No, I can't** consider your request **because** I don't have everyone's schedule." "**No, I can't** help you **because** I've got an urgent project to complete."
>
> Turn your negative responses into positive ones: "**Sure**, I will be glad to consider your request, **as soon as** I know everyone's schedule," makes it a positive response. Often, we <u>can</u> give people what they want. Perhaps not immediately, but eventually.

Putting it into Words: Occasionally, a doctor encounters 'difficult' patients who require that extra measure of verbal sensitivity. Below are sample responses to various types of patients whom doctors are likely to meet in their practices. The words are so well put together that I am quoting the entire text *verbatim*. This came from a handout that was originally written by Dr. Martin Lipp of the Gilroy Medical Offices (Kaiser Health Plan).[38]

Angry Patient

- I'm glad you're telling me about this.
- If that is what you think (or perceive), your feelings are certainly understandable.

---

[38] Permission to reproduce this section was kindly given by Dr. Martin Lipp.

- I understand. Lots of people would be upset in similar circumstances.
- If that's what you feel, I'm glad you're not keeping your feelings inside.

Complaining Patient

- It's frustrating when things don't go the way we would wish.
- I appreciate your telling me the way you see it.
- If that's the way you see it, I am glad you are speaking up.
- Observations like yours keep us on our toes; thank you for bringing that to my attention.

Laundry List Patient

- It's frustrating to have a whole list of concerns when your doctor can only deal with a few of them.
- I appreciate that you have organized your concerns — an organized patient can be a big help to me.
- I'm glad you wrote everything down — let me see your list.
- When there is a lot going on, sometimes the only way to keep track of everything is to make a list.

Knowledgeable Patient

- You really know a lot about this — how have you managed to learn so much?
- Your knowledge makes it possible for the two of us to approach this problem as a team.
- A knowledgeable patient is a wonderful ally for a doctor — I'm glad you read up on this topic.
- You obviously know a lot about this topic — perhaps we can teach one another.

Dependent Patient

- It would be really helpful for you to have someone to lean on now, wouldn't it?

- It's awfully rough having to go through this by yourself, isn't it?
- I'm glad you are telling me this. It helps me to understand your situation so much better.
- It must be frustrating for a self-sufficient person like you to be needing someone's help at this time.

Manipulative Patient

- If that's what you think you need, I'm glad you asked.
- From your point of view, it certainly makes sense to ask for that.
- If you see me as an obstacle to something you want, it's certainly understandable for you to be upset.
- Since I don't feel comfortable giving you what you want, let me help you consider other alternatives.

Pain Medication Patient

- When you are in pain, it certainly makes sense to ask for medication.
- If you think medications are the only thing that will help, I understand your wanting them so badly.
- It's frustrating for you to want me to prescribe medication and for me to say no.
- Your wishes are perfectly understandable; but since I don't feel comfortable writing this prescription, let's consider how we can work this out together.

Kaiser (Health Plan) Critic

- I can understand your unhappiness if you feel this health plan isn't meeting your needs. However, my job is to help you as best I can within the plan's framework.
- I understand what you are saying, and I am glad you spoke up. I encourage you to speak with Patient Assistance about this. However, you and I need to get beyond this so we can save time for your medical problems.
- If that's how you feel, it's good you are speaking up. I want to do the best job I can of taking care of your health. Are your criticisms about the health plan or about me in particular?

• I appreciate your having told me your concerns. Why don't you talk to the Health Plan Rep about your criticisms of Kaiser, and the two of us can use our limited time together to focus more on your medical concerns.

'The Four Habits of Highly Effective Clinicians': Drs. Richard Frankel and Terry Stein produced an excellent handbook in 1996 entitled *The Four Habits of Highly Effective Clinicians: A Practical Guide* that focuses on effective communication with patients.[39] The handbook was produced for the Kaiser Permanente Northern California Region, where Dr. Stein works as the director of clinician-patient communications. The four habits are:

(1) Invest in the beginning
(2) Elicit the patient's perspective
(3) Demonstrate empathy
(4) Invest in the end

Habit 1. Invest in the Beginning: The goal here is to establish rapport and build trust rapidly, thereby improving diagnostic accuracy. Physicians will need to develop skills that: (1) Create rapport quickly; (2) Draw out the patient's concerns; and (3) Plan the visit with the patient. The payoff is faster access to reason for the visit, better diagnostic accuracy, and less work. It also facilitates negotiating an agenda, and it prevents the patient's "Oh, by the way" complaint as he or she is about to leave the office.

Habit 2. Elicit the Patient's Perspective: This facilitates the effective exchange of information, which improves patient compliance. The physician is expected to: (1) Ask for the patient's ideas; (2) Elicit specific requests; and (3) Explore the impact on the patient's life. Patients will often give you the diagnosis, hidden concerns will surface, and alternative treatments and requests for tests will be revealed. A missed diagnosis of depression and anxiety is less likely, and this habit encourages patients to take a more active role.

Habit 3. Demonstrate Empathy: Empathy is the action of understanding, being aware of, being sensitive to, and vicariously experiencing the feelings, thoughts, and experiences of another. It enables the clinician to hear

---

[39] ©1996, Physician Education & Development, Kaiser Permanente Northern California Region. Reproduced with permission kindly provided by Dr. Terry Stein.

what the patient is <u>not</u> saying. Empathy recognizes, identifies, and acknowledges the patient's feelings. Empathy is not sympathy, which often denies or discounts feelings, thus diverting the focus away from the patient. Empathic responses use words like: *"You must be concerned"; "That must have been a shock for you"; "This must be a difficult time for you"; and "It sounds like you have many responsibilities."* Look for patient responses like *"You got it, Doc"; "Exactly"; or "That's how I feel."*[40]

On the other hand, words of sympathy include: *"I know how you feel. I've had it myself"* (takes the focus off the patient; instead of being a listener, the doctor is a speaker); *"Don't feel bad, everyone has problems sometimes"* (denies feelings); *"Don't worry about it; it'll work out"* (discounts feelings); and *"At least you can recover from this"* (shuts out further questions).[41]

The goal of empathy is to show caring and concern. This results in patient satisfaction. Be open to the patient's emotions, make at least one empathic statement, convey empathy non-verbally, and be aware of your own reactions. The payoffs are that the patient feels understood and becomes more trusting, which leads to better diagnostic information, better adherence, and better outcomes.

Habit 4. Invest in the End: This increases the likelihood of adherence and positive health results, with corresponding clinician satisfaction. It requires developing the skill of delivering diagnostic and treatment information in a simple, understandable fashion, and of providing patient education. One can expect a positive impact on health outcomes and compliance, and a reduction in return calls and visits.

Wall Street has taken notice of Kaiser's 'Four Habits' program, calling it 'a good road map for patients who want to get the most out of their doctor's visit.'[42]

---

[40] Coulehan, JL et al. "Let Me See If I Have This Right . . . ": Words That Help Build Empathy. *Ann Int Med* 2001; 135:221–7.
[41] Taken from a newsletter of the Hawaii Medical Library.
[42] Landro, L. Teaching Doctors How to Interview. *The Wall Street Journal*, September 21, 2005, p. D5.

DOCUMENTATION

Lawsuits may be filed months or years after the incident. In one study, the average time between incident and trial in metropolitan areas was 67 months. The doctor's recollections are likely to be vague through the passage of time. There is the old adage that *'What was not documented was not done.'* Or: *'Good records, good defense; bad records, bad defense; no records, no defense.'* Should a claim be made, you will be set back if the relevant information is not documented. Well-kept detailed records will provide the best support. Besides, good records are vital for quality healthcare, especially in complicated cases.

Health professional records tend to be inadequate and illegible. Instead of helping, these records may actually hurt the practitioner. Document all discussion and procedures clearly and completely, especially if there was no specific consent form. If an inadvertent complication should occur, keep detailed notes. Be factual, non-judgmental, unemotional, and accurate. Do not describe the patient or staff in harsh terms even if he or she is partly or mostly to blame. The records should reflect the doctor's caring attitude and a primary concern for patient well being and safety.

Template charting is becoming popular as a time-saving measure. The doctor merely makes a check-mark alongside items on a pre-printed 'record.' Be sure that what is checked corresponds accurately to what transpired between the doctor and the patient. Wholesale checkmarks, made indiscriminately, will seriously undercut the doctor's credibility and may lead to allegations of fraud.

Never ever, alter the medical records. Such alteration is frequently taken to mean that the doctor has something to hide, and will be used by the plaintiff attorney for maximum effect. If an honest error is made in the charting, place a single cancellation line across the error, so that the error can still be read, and enter the new correct information below or next to the earlier entry. Do not erase, 'white-out' or otherwise obliterate the erroneous entry. Deliberate alteration of the medical records to cover up is unethical. In the movie 'The Verdict,' Paul Newman plays an alcoholic attorney who won a malpractice lawsuit after the defendant was caught altering the medical records.

HOW TO HANDLE A WRITTEN COMPLAINT

Treat all complaints seriously and do not ignore them. Written complaints or queries require a prompt written response but consult your insurer before taking any action. In drafting a response letter to a patient or a regulatory agency, the doctor should be guided by the following:

(1) **Be honest:** The truth usually surfaces in due course. Candor and trustworthiness are virtues expected of all doctors.
(2) **Be accurate:** Review carefully the medical records that pertain to the incident, with compulsive attention to factual accuracy.
(3) **Be focused:** Do not ramble on and on regarding unrelated or tangential issues. Focus on what the patient is complaining about. No one is interested in your views regarding your philosophy of medical practice or the healthcare system.
(4) **Be brief:** Your main points will be lost, or go unread, if your reply is 10 pages long.
(5) **Be professional:** Your letter may be reproduced at a later hearing or in court, and it should not come back to haunt you. Imagine the embarrassment to see a letter that is barely legible, replete with grammatical errors, or just infantile in tenor. If you cannot write well or effectively, get help from someone who can.
(6) **Be humble:** Arrogance at this stage (or at any stage for that matter) may prove disastrous. Adopt a contrite and humble tone. Blame no one, especially the patient.
(7) **Be a patient advocate:** Show that patient well-being is always your first and last concern.

FACTORS THAT PROMPT LITIGATION

The American litigious mindset has spawned lawsuits of every imagination. This is true for construction disputes, personal injuries of all types, and suits against teachers, priests and even parents. One commonly cited explanation is our legal system with its contingency fee system and runaway jury verdicts that occasionally hit the millions and make big splashes in the media.

However, it is generally accepted that far fewer lawsuits are filed than there are malpractice events. Some put this ratio at as low as, or lower than, one in ten. In other words, most injuries that result from substandard treatment go unchecked by the legal system. Undoubtedly, many patients do not make a claim because of ignorance and lack of sophistication over the legal and/or the medical system.

On the other hand, many patients actively decide to forgo filing a lawsuit. Why do some patients sue and not others? A study in 1992 by Hickson and colleagues identified the factors that prompted families to file malpractice claims.[43] They interviewed 127 mothers of infants who had experienced permanent injuries or death. The reasons given for filing the lawsuits were:

Advised by knowledgeable acquaintances 33%
Recognized cover-up 24%
Needed money 24%
Recognized that child would have no future 23%
Needed information 20%
Decided to seek revenge or protect others 19%

Families also expressed dissatisfaction with physician-patient communication, asserting that physicians would not listen (13%); would not talk openly (32%); attempted to mislead them (48%); and did not warn about long-term neuro-developmental problems (70%).

Studies on the psychology behind a malpractice lawsuit have unveiled several consistent factors that prompt litigation. These factors center on poor communication and the attitude of the doctor and/or healthcare institution. Physicians who lack people skills or who are just too busy to spend time talking to patients are the ones most apt to get themselves in trouble. Some plaintiffs equate a bad outcome with bad care. Others may have totally unrealistic expectations (case in point: cosmetic surgery). Although these claims are relatively easy to defend, many of them could have been prevented in the first place if the doctor had taken time to explain both before and after the event.

Here's a check-list of the characteristics of doctors who receive the most patient complaints and are involved in so called 'risk-management' events:

(1) Male doctors
(2) Surgeons

---

[43] Hickson, GB et al. Factors that Prompted Families to File Medical Malpractice Claims Following Perinatal Injuries. *JAMA* 1992; 267:1359–63.

(3) Doctors with poor bedside manners
(4) Busy doctors

The nature of the injury and the quality of the doctor-patient relationship are by far the most important factors prompting a lawsuit. Patients rarely sue doctors whom they like, and are amazingly forgiving of errors if an apology is offered and there are plans to prevent a similar error in the future. Two factors that do not appear to make a difference: Age and country of medical training.[44]

Anger: <u>Question:</u> *What do you call an angry patient?* <u>Answer</u>: *A plaintiff.*

This is a familiar refrain of malpractice attorneys and should serve as a warning to doctors. Anger lies at the root of all malpractice claims. It is the emotion that prompts the patient or family to seek legal redress. The anger may not originate entirely from the adverse result itself. It may be the mistreatment of the patient or family in the form of a negative, indifferent, or arrogant attitude of the healthcare provider or institution. Unjustified delays, stonewalling, not returning calls, and refusing to meet with the aggrieved, are all warning signals. These attitudes betray a lack of caring, which angers the injured patient who may then seek legal redress. And angry jurors who may have had similar bad experiences are the ones most likely to return a runaway verdict.

When patients and family members are initially surprised then disappointed and upset at the results or the way they are treated, they may start to consider seeing a lawyer. A prominent plaintiff attorney who gets 150–200 calls each month from patients who want to sue their doctors or the clinics noted that the majority of these calls were prompted by poor communication. The patients may not have a serious injury or a meritorious claim, but they are so angry about their physician's inability or refusal to communicate that they contact an attorney to vent their anger. They are looking for revenge. In one study of 263 patients and families which had successfully sued a physician, more than half said that they wanted to sue him even <u>before</u> the alleged malpractice incident took place.

---

[44] *Medical Economics*, July 11, 2003.

Experienced risk managers and attorneys have pointed out that there are other emotions one must understand, since they can lead to anger. These emotions include frustration, guilt, and embarrassment. Frustration commonly results from long waits before being seen and for appointments and prescriptions. Guilt feelings may occur when a patient or family member perceives that he or she could have prevented the accident, illness, or injury to self or loved one. It is natural to want to pass on the blame to someone else, which may be the healthcare provider. Finally, gossip and breach of confidentiality will embarrass most patients. If unresolved, embarrassment can turn into anger, and we now have the seeds of a malpractice complaint.[45]

Poor Service: It has been pointed out that at Disney theme parks, everyone is a guest. Only people with a smile on their face are hired. Hospitals, on the other hand, may hire unskilled personnel who may receive little or no training before being put on the job. Overworked staff frequently treat patients as burdens, and oftentimes no one is present who is empowered to promptly address customer complaints.

We are urged to think of Nordstrom's whenever we think of service. This retailer is a legend when it comes to putting the customer first. The story is told of the man who asked for a refund on a car tire that had obviously been in use for some time. Instead of arguing over whether the tire failed to meet the customer's satisfaction, or whether it was defective, the sales clerk promptly and smilingly refunded the full cost. Nordstrom's gained a happy customer, even though it does not sell tires.

Disparaging Remarks or Attitudes: Some lawsuits are filed because of suspicion raised in the patient's mind by disparaging remarks made by one practitioner about another's diagnosis or treatment. An example: *"Who did that to you?"* (doctor or nurse pointing to a large abdominal surgical scar). Without having all the facts, one should avoid making critical comments about a colleague's work. Worse yet, an unthinking healthcare provider sometimes enters disparaging remarks into the medical records. Statements like *"physician refused to respond"* should be replaced by *"no response yet, will try again."* Disagreements between doctor and nurse require proper channels

---

[45]Some of the materials in this section (Factors that Prompt Litigation) have been excerpted or adapted from an excellent videotape entitled "Seven Secrets of Avoiding Medmal Suits." Videotape, produced by Stanstead Multimedia Corp., Rockford, Illinois, copyright 1996, and presented by Frew Consulting Group, Ltd.

for resolution. The medical record is <u>not</u> one of them. In the event that a staff member makes a mistake, a physician should respond in a calm and professional manner, without actively criticizing the person in front of the patient or family.

The staff can also be the cause of the lawsuit. Staff members are typically the first people in contact with the patient and family. They must listen to the patients and make them feel important. Staff should be as well trained in customer service as in quality care. One should counsel, and if necessary, discipline those who engage in gossip.

Money Disputes: Out-of-pocket costs for medical care are skyrocketing, and patients are less tolerant if they perceive insufficient value for their money. Because of the increasing impersonality of the medical encounter and the commercialization of healthcare, we can expect injured patients to retaliate, especially if they feel unjustifiably overcharged.

Aggressive billing practices run the risk of further angering a patient who may have suffered an adverse outcome. Great tact and sensitivity is necessary to avoid turning this into a malpractice claim. This is not to say that only good clinical results deserve to be paid for. However, a patient angry over the bill may have an unspoken agenda or otherwise legitimate reason for feeling hostile.

Some doctors resort to a collection agency for unpaid patient bills. My own experience and observation is that this usually does not amount to much, and may be counter-productive. The habit mostly benefits the debt-collecting agency. Consider adopting the policy of writing off all unpaid bills in the name of public service, although this may be impractical for institutions or where the sums owed are large.

PATIENT SATISFACTION

Customer satisfaction is a familiar refrain in the general marketplace. If there is one good reason to refer to patients as 'customers,' it is the hope that perhaps this will prompt the healthcare provider to be more cognizant of their service needs. All practitioners should find out from their patients whether they are satisfied with the care they are receiving. It is an error to assume that because patients keep returning to see the doctor that they are satisfied with the service. Other reasons may apply, e.g., convenient location, familiarity, embarrassment in switching doctors, and so on.

The practitioner who is late for appointments should have the clinic assistant relay the information, with apologies, to the waiting patients. Studies show that patients are willing to tolerate a wait of 15–30 minutes in the doctor's office before becoming annoyed and angry.

The mantra of 'patient (client) satisfaction' is sounding louder within Medicine's new knowledge and speed-based transformation. For example, The American Board of Internal Medicine has recently come up with a 10-item patient assessment questionnaire to help evaluate continuous professional development. This is part of the Board's re-certification program, which all internists will be obligated to complete every 10 years. Patients are asked to anonymously evaluate their doctors via a telephone response system. Below are the sample questions.

How is your doctor at:

(1) Telling you everything
(2) Greeting you warmly
(3) Never talking down to you
(4) Listening carefully
(5) Showing interest in you as a person
(6) Warning you during the physical exam about what he is going to do
(7) Discussing options with you and asking your opinion
(8) Encouraging you to ask questions
(9) Explaining what you need to know
(10) Using words you can understand

This is a good checklist to apply in one's practice. Doctors working in large groups or clinics should be even more attentive to these patient expectations, as the doctor-patient relationship may be at greater risk of becoming impersonal. Such groups or clinics should address additional issues, including patient privacy, staffing, waiting-time, and overall warmth and efficiency.

SUMMARY POINTS

HOW TO AVOID BECOMING A HORROR STORY

- Medical malpractice lawsuits are often as much about poor communication and patient or family anger as they are about poor quality of care.

- Prevention is the best defense. This means developing both clinical and communication skills.

- Put the patient first. Competent care assures patient safety and is sensitive to patient satisfaction.

- Words can trigger embarrassment, suspicion and anger. Use words with care and think before speaking.

- Good record keeping and a competent defense counsel are the best allies when facing a lawsuit.

21

Medical Errors and Their Disclosure

LEARNING FROM THE AVIATION INDUSTRY[46]

On the fateful night of October 31, 2000, Singapore Airlines SQ 006 taxied onto a closed disabled runway at Taiwan's Chiang Kai-shek International Airport, and crashed into construction equipment as it took off.[47] It was Singapore Airlines' first aviation tragedy, and it involved three pilots whose past record was unblemished. The accident claimed 83 lives.

The facts of the accident are incontrovertible. SQ 006 took an erroneous turn onto runway 05R that had been closed for repairs. Poor visibility, the absence of barriers, the unexplained presence of lights on the closed runway, and the all-clear signal from control tower were factors that misled the pilots into making the tragic turn. In its formal accounting released on April 27, 2002, Taiwan blamed the accident squarely on the errant pilots and bad weather, and relegated airport deficiencies to a 'risk' category rather than as causative or contributory factors. It recalled the three pilots to Taiwan to face further questioning and the specter of criminal prosecution. Singapore had suspended the pilots after the tragedy, but drew quite different conclusions from these same facts. Singapore Airlines assumed full responsibility for the accident ("it was our pilots and our plane"), but the Singapore Transport Ministry refused to blame any single person or factor, preferring to call it a system error. In the name of passenger safety, it recommended learning

---

[46] Excerpts of this chapter has been previously published in the *S Med J* 2002; 43(6):276–8 and is reproduced with the kind permission of the Editor.
[47] *The Straits Times*, November 1, 2000.

from the cumulative errors, and correcting all deficiencies so that accidents of this type will never happen again.[48]

This aviation tragedy holds important lessons for the medical profession. Just as pilots are entrusted with the serious responsibility of ensuring the safety of the flying public, doctors are expected to safeguard the well-being of their patients. Just like aviation, the healthcare system is complex with many opportunities for mistakes. There is now widespread recognition that medical errors are responsible for many hospital injuries and deaths. Lucian Leape highlighted this problem many years ago,[49] and a recent report from the Institute of Medicine entitled: *'To err is human'* has brought the matter to public prominence. The report places medical errors as the cause of between 44,000 and 98,000 annual fatalities, which makes it the fourth most common cause of death.[50]

The Institute of Medicine has proposed nine broad recommendations to reduce medical errors, including establishing a nationwide mandatory reporting system that provides for the collection of standardized information by state governments about adverse events that result in death or serious harm. Its mantra is for the healthcare system to shift from a culture of blame to one of patient safety. In its follow-up report, *'Crossing the Quality Chasm: A New Health System for the 21st Century,'* the Institute called on Congress to create a $1 billion fund to support projects targeting safe, effective, patient-oriented, timely, efficient, and equitable patient care.

JCAHO has gotten into the act, and has made patient safety its overriding criterion in its accreditation process. It has enunciated seven patient safety goals for healthcare organizations. These are:

1. Improving the accuracy of patient identification
2. Improving the effectiveness of communication among caregivers
3. Improving the safety of using high-alert medications
4. Eliminating the wrong-site, wrong-patient, wrong-procedure surgery
5. Improving the safety of using infusion pumps

---

[48] *The Straits Times*, April 27, 2002.
[49] Leape LL. Error in Medicine. *JAMA* 1994; 272:1851–7.
[50] Institute of Medicine: To Err is Human: Building a Safer Health System. National Academy Press, Washington D.C., 2000. The full text of the report is available online at www.nap.edu/readingroom. The Institute of Medicine's home page is at www.iom.edu.

6. Improving the effectiveness of clinical alarm systems, and
7. Reducing the risk of healthcare-acquired infections.[51]

MEDICAL ERRORS

The term medical error denotes a preventable adverse event, which in turn is defined as an injury caused by medical management rather than the underlying condition of the patient. It is not the same as negligence (see Chapter 2: What Is and Isn't Malpractice). The current approach to preventing medical errors is to assign individual blame rather than look at them as a problem with the system. Despite the notion that healthcare professionals are not supposed to make mistakes, the truth is that they often do. Fortunately, the majority of medical errors cause no serious harm. Studies published by Harvard researchers in 1991 indicated that 3.7% of hospitalized patients suffer significant iatrogenic injuries, typically from errors or negligence.[52] According to Leape, we make an average of 1.7 mistakes per patient per day in the intensive care unit. To be sure, almost 200 patient-care activities take place daily in the intensive care unit. Still, Leape makes the point that a 99% level of proficiency, i.e., a 1% failure rate, is too high to be tolerated in a hazardous industry like ours. At 99.9%, there would be two unsafe plane landings at O'Hare airport each day; the U.S. post-office would lose 16,000 pieces of mail; and 32,000 bank checks would be deducted from the wrong accounts every hour.

Doctors respond predictably to medical errors. They deny them, hide them, and even attempt to bury a few of them. They typically become defensive, and blame others for the mistake — the nurse, the hospital, even the patient. The doctors most deserving of support are the ones who suffer in silence, fearing discovery and publicity, depressed with guilt and fallen esteem over what is perceived as failed duty.

Society, in conspiracy with the profession, has perpetuated the myth that good doctors do not make mistakes. Voltaire in 1764 compared doctors to God: *"They even partake of divinity,"* he wrote, *"since to preserve and renew is almost as noble as to create."* During postgraduate residency training, all program directors will exhort their trainees to strive for perfection.

---

[51] Available at www.jcaho.org, Accessed June 2004.
[52] Brennan, TA *et al.* Incidence of Adverse Events and Negligence in Hospitalized Patients: Results of the Harvard Medical Practice Study I. *N Eng J Med* 1991; 324:370–6.

Unfortunately, this is an unrealistic goal, as one cannot escape making at least a few mistakes despite the best of intentions and the highest competence.

It has been estimated that during any overseas commercial flight, a human error or an instrument malfunction occurs every four minutes — yet each event is promptly recognized and corrected.[53] This is the science of system errors and failures at work, and it can help the healthcare industry. Better standardization, task design, checks and counterchecks, systems monitoring and backup, and automatic alerts will go far in reducing errors in hospitals and clinics. There is proof to support this approach. The death risks in anesthesia, for instance, have been dramatically reduced in the past two decades because of monitoring devices such as the pulse oximeter and capnograph. Unit dosing in the hospital pharmacy is another innovation that has reduced medication errors. In the near future, electronic orders will do away with illegible handwriting, a key cause of prescribing and dispensing errors.

However, it is of utmost importance for physicians to learn from their mistakes, and be able to identify and tabulate them. This will likely not happen in an atmosphere of fear. The purpose of reporting must therefore be to educate, not punish; restore, not denigrate. A model that focuses on fair compensation and improvement in healthcare standards must replace our fault-based malpractice system. The profession should encourage its senior members and clinical teachers to share their adverse experiences with junior colleagues. It is an effective way of saying, *"We all make mistakes — let's learn from them to benefit our patients."*[54]

Being smart and diligent, even vigilant, may not be enough. Dr. David Gaba, an anesthesiologist at Stanford, has emphasized that safety measures should focus on process rather than people. He has advanced criteria that define 'High Reliability Organizations' — systems that are virtually failure-free in extremely hazardous environments. They include: (1) Optimal organizational structures and procedures; (2) Intensive training during operations and simulations; (3) Creating and maintaining active cultures of safety; and (4) Maximizing learning from incidents and accidents.[55] Sadly, Dr. Gaba

---

[53] Leape, LL. Error in Medicine. *JAMA* 1994; 272:1851–7.
[54] Tan, SY. When Doctors Make Mistakes. *Haw Med. J* 1996; 55:135, 146.
[55] Gaba, DM. Structural and Organizational Issues in Patient Safety: A Comparison of Health Care to Other High-Hazard Industries. *California Management Review* 2000; 43:83–102.

laments that *"while healthcare contains seeds of each of these approaches, and some of the seeds are sprouting, there remains a long way to go."*[56]

WHEN AN ERROR HAS OCCURRED

Patients who trust and think well of their doctors rarely sue them, even if they sustain bad results. They are willing to forgive and accept errors, oversight and carelessness. This truism simply attests to the overarching power of the doctor-patient relationship. When a mistake has been made, the best approach is honesty. Patients have a right to be told the truth rather than to have the facts hidden from them, or worse yet, lied to. Such forthrightness, shared with compassion and humility, is more likely to prevent than to precipitate a lawsuit.

This approach is at odds with the traditional legal advice to say and admit nothing. However, recent studies are beginning to recognize that honesty is the best policy. A humanistic risk management policy has been operational at a Veterans Affairs medical center since 1987. The protocol included early injury review; steadfast maintenance of the relationship between the hospital and the patient; proactive disclosure to patients who have been injured because of accidents or medical negligence; and fair compensation for injuries.[57] In an accompanying editorial, this anecdote was attributed to an attorney:

"In over 25 years of representing both physicians and patients, it became apparent that a large percentage of patient dissatisfaction was generated by physician attitude and denial, rather than the negligence itself. In fact, my experience has been that close to half of malpractice cases could have been avoided through disclosure or apology but instead were relegated to litigation. What the majority of patients really wanted was simply an honest explanation of what happened, and if appropriate, an apology. Unfortunately, when they were not only offered neither but were rejected as well, they felt doubly wronged and then sought legal counsel."[58]

---

[56] Gaba, DM. Commentary, *The Pharos*, Winter 2002, pp. 10–11.
[57] Kraman, SS and Hamm, G. Risk Management: Extreme Honesty may be the Best Policy. *Ann Int Med* 1999; 131:963–7.
[58] Wu, AW. Handling Hospital Errors: Is Disclosure the Best Defense? *Ann Int Med* 1999; 131:970–2.

Honest mistakes should be disclosed in a timely and compassionate manner. The best way to arouse suspicion and anger is to stonewall a patient's inquiries. Admission of errors is not necessarily the same as admitting or accepting fault. Saying *"I'm really sorry this complication occurred"* is not the same as saying *"I'm sorry I made a mistake and it's my fault that there has been a complication."* At times, however, it may well be appropriate to express outright responsibility for an error. Choosing the right words is important, as is the physician's attitude and body language.

To avoid disclosure is to harm the physician-patient relationship by injuring trust. When we disclose medical errors, many patients may actually be aware that such a disclosure is difficult for us and appreciate our willingness to put their welfare above our own professional or financial interests. Plaintiff attorneys know that their clients are more likely to pursue punitive legal action when they suspect a cover-up.[59]

The attending physician should be present and act as the primary speaker. Perhaps other members of the medical staff, such as the medical director or other senior staff physician, could attend to reinforce that an investigation is ongoing. An attorney should <u>not</u> be present because it may send a message that we anticipate a lawsuit, and such presence may shift the interaction away from a trusting, benevolent physician-patient relationship toward an adversarial one. In many hospitals it is the risk manager or patient advocate who meets with the patient and family to explain the circumstances. As well-trained and sensitive as these individuals may be, the attending physician remains the one in the best position to disclose the error — assuming he or she has the right attitude and training for the task.

There have been publications on the breaking of bad news, of which disclosing a medical error is a subtype.[60] However, disclosure of an error is different for a number of reasons. For one thing, the doctor rightly fears a confrontation and the possibility of a lawsuit. For another, there is the potential

---

[59]American Society for Healthcare Risk Management. Disclosure of Medical Errors: Demonstrated Strategy to Enhance Communication. Panel Discussion and Video Broadcast. Chicago, IL. June 20, 2001.
[60]Iserson, KV. *Grave Words: Notifying Survivors about Sudden, Unexpected Deaths.* Galen Press, Ltd. Tucson, AZ, 1999; Buckman, R and Kason, Y. *How to Break Bad News: A Guide for Health Care Professionals.* The Johns Hopkins University Press, Baltimore, MD, 1992; Maguire, P. *Communication Skills for Doctors: A Guide to Effective Communication with Patients and Families.* Oxford University Press. New York, NY, 2000.

erosion of the doctor-patient relationship. In order to be effective, physicians need to learn and rehearse the actual words to use when discussing errors with patients and family.

BASIC TASKS OF DISCLOSURE[61]

How do patients want their doctors to handle mistakes?[62] They will have strong emotions, including anger and a tendency to blame. They may lose trust in their physician and all medical personnel. They may attempt to punish those they feel are responsible for the error. Patients differ in age, educational, socioeconomic and cultural backgrounds, so we must take these factors into account. They all want to know the details of the error and the circumstances leading to it. But all of the facts may not be readily available at the time that the error first becomes apparent. The patient may have a natural desire to know every detail immediately, and we should refrain from speculating as to the cause or who might be the responsible individual. It is more acceptable to state that we don't yet know all of the facts and details. This must come across not as a smokescreen, but as a statement of fact.

Patients will also want to know what this error means to them. Will additional treatment be necessary? Will there be complications? How many more blood draws will they need, or radiological studies, or days in the hospital? The physician should discuss these consequences, and if warranted, assure the patient that there will be no additional financial costs to pay.

When an error occurs, it is natural to expect an apology. From the time we enter school as children, we are taught that the right thing to do if an accident happens is to go to the person wronged and apologize. An apology is appropriate in the context of a medical error as well. A phrase often heard in a plaintiff's attorney's office is: *"That doctor never even said he was sorry!"* Apologizing for an error, even if we are apologizing on behalf of our system, is the right thing to do.

It is possible that the patient may feel that since the hospital or the doctors have so much to lose, a cover-up is likely. We must promise to continue to be

---

[61] Dr. Jason Fleming who recently finished an Emergency Medicine residency, assisted in the preparation of this section during a senior medical student elective.
[62] Witman, AB *et al.* How Do Patients Want Physicians to Handle Mistakes? A Survey of Internal Medicine Patients in an Academic Setting. *Arch Int Med* 1996; 156:2565–9.

truthful. Giving our assurance that we will continue to be truthful is a simple way to buttress our physician-patient relationship at this crucial time.

There are, therefore, four basic tasks when disclosing errors. These are shown in Table 10, and are discussed in greater detail in the illustrated case scenarios that follow:

From the first days of clinical training in medical school, we are taught that the non-verbal aspect of our communication with patients is critical in determining the outcome of the interview. Our appearance, posture, and gestures will define us in the minds of our patients. Before stepping into the interview, it is useful to take a few moments to prepare the appropriate messages we wish to send to the patient.

Throughout the disclosure process, it may be useful to employ a stop-and-go strategy where a little information is given in a gentle way at first, followed by a pause to allow the patient to adjust to this new information. The patient then has an opportunity to actively ask for further information and the physician will have immediate feedback on the appropriate pace of the disclosure. This method has been widely used to deliver bad news, and should be equally applicable to the disclosure of errors.[63]

Step 1 — The Details: Our first task is to give the patient all the details of the circumstances that contributed to the error, as we understand them at this time. Included in this step is the actual disclosure that a medical error has occurred. This is an uncomfortable moment for everyone, including the physician. It should be remembered that the person most uncomfortable is

Table 10: The Four Tasks of Disclosure

1. All the **DETAILS** we have at this time
2. What it **MEANS** to the patient
3. **APOLOGIZE** that the error occurred
4. Our **PROMISE** to continue to be truthful

---

[63] Desmond, J and Copeland, LR. *Communicating with Today's Patient: Essentials to Save Time, Decrease Risk, and Increase Patient Compliance.* Jossey-Bass, Inc. New York, NY, 2000. See also footnote 60.

the patient, because of the powerlessness and uncertainty involved in being a patient in a complex medical environment.

Utilizing the aforementioned stop-and-go method of gentle disclosure, begin by using a phrase such as: *"I do not have good news,"* or *"Something very troubling has happened."* Pause and allow the patient to indicate that they are willing to have more information.

Continue with a statement that provides more details, such as: *"It seems that there was a mistake with your medication."* This statement communicates that the nature of the bad news is an error, what the error was related to, in this case, medication, and most importantly it demonstrates candor on the part of the physician. Avoiding the use of technical or legal language strengthens the patient's trust that you are not hiding anything.

Patients may be demanding in their desire to know more details, and may even subtly attempt to discover who is at fault. This is a natural reaction, but the physician must not feel compelled to enter into blaming. It is acceptable to say, *"We just don't have all the details yet. I know you need more information, but I would hate to tell you something that isn't correct."* Giving misinformation at this crucial stage will increase confusion later, and may even appear as though a cover-up was attempted. Avoid speculation. Follow up by saying that we are searching for the answers, such as: *"We are making every effort to find out more information,"* or *"The hospital is investigating this and finding out the answers is our top priority."* In this way, the physician will show that steps are being taken to discover the cause of the error, and that answers are important to the hospital and staff and the patient.

Step 2 — What it Means to the Patient: Patients need to know what the consequences of the error will be for them. Interfacing with modern healthcare is frightening for patients because they don't know what to expect. Any further tests, procedures, hospital days, or other medical intervention must be included in the disclosure. This will also help the patient to appreciate the severity of the error. If they need to expect *"a simple blood test to check your liver,"* they may understand that the error is not as serious as if they hear *"we will need to monitor your liver function closely over several days in the hospital."* By telling patients what to expect, we assist their psychological adjustment to the new situation and decrease their anger by communicating that we are anticipating what they want to know.

Step 3 — The Apology: There have been differing opinions as to whether physicians should say the words *"I'm sorry"* during the interaction with the patient. Some suggest that doing so will give ammunition to plaintiff's attorneys who may try to prove culpability by showing the physician used those words. Some hospital attorneys have suggested phrases such as *"We deeply regret that this occurred"* or *"You have our deepest sympathy"* in place of *"I'm sorry."*

Trying to avoid saying *"I'm sorry"* leads to awkward and stilted language that no real person uses in day to day speaking, especially to one they care about. Saying *"I'm sorry"* is a natural and human expression of empathy, and should not be avoided due to fear of potential litigation. Being emotionally vulnerable demonstrates caring, and may decrease the patient's tendency to blame. Saying, *"I'm very sorry this happened. I feel terrible about this,"* shows humanness and empathy. Caveat: Saying *"I'm sorry"* is not the same as baring one's soul and making damaging statements and hasty admissions of fault before all the facts are known. Defense attorneys regularly warn against this type of hasty admissions of fault as they can never be retracted from the patient's memories — or the juror's ears.

Step 4 — The Promise to be Truthful: At the close of the meeting, we must combat any potential perception of a cover-up by committing to continued truthfulness. Using everyday language, a statement such as *"I want you to know that no matter what, I'm going to make sure you get all of the information as we go forward"* will show that you are the patient's ally. Another sentence might be: *"I will do my best to keep you up-to-date on everything that is going on."* And always end the interview with: *"Do you have any questions you wish to ask of me at this time?"* It shows openness, candor and honesty.

Words and Phrases to Avoid: One should be aware of potentially harmful statements and avoid them during the meeting. Iserson offers a list of such phrases during the notification of sudden deaths.[64] Many of the same principles can be applied to the disclosure of errors.

Whereas *"I'm sorry this happened to your mother"* is fully appropriate, one should not use words like: *"I'm sorry we did this to her"* or *"I'm sorry*

---

[64] Iserson, KV. *Pocket Protocols — Notifying Survivors about Sudden, Unexpected Deaths.* Galen Press, Ltd., Tucson, AZ, 1999.

we gave her the wrong treatment." Nor should one speculate on the blameworthiness of members of the healthcare team. Thus, avoid words like: *"The call-in nurse made a mistake ... "* or *"The intern is new and unfamiliar with the procedure ... "* What appears to be the cause of the error at the time of discovery may not be the true causative event, and if differing stories are given at different times, the suspicion of a cover-up is inevitable. In assiduously avoiding the blame game, be mindful that one should not point a quick finger at an easy target — the hospital or clinic. Do not be tempted to use words like *"They are always understaffed"* or *"I've told them about this potential problem in the past."*

Balancing reluctance to speculate against the appearance of withholding information is a difficult task for the disclosing physician. The solution lies in explaining why you don't know or can't answer the question immediately. The patient may be frustrated and angry. This is a natural response, and should be accepted with patience and caring. The commitment to truthfulness is a crucial message to reiterate.

Follow-up Discussion: Once all the facts are known, we should promptly inform the patient of what exactly happened and why. Share the information in a factual and empathic manner without directly blaming anyone by name. Preferred words are *"one of our staff members, one of our nurses or doctors."* Together with the apology, we should reassure the patient of the steps taken to prevent this type of error from happening again in the future, so no one else would be harmed.

SCENARIOS AND THE WORDS TO GO WITH THEM

In addition to choosing the right words, the demeanor and body language of the physician are obviously equally important.

Scenario 1: Harmful Mistake was Unavoidable

A 5-year-old child presents to the local Emergency Department with fever for one day. Workup revealed mild dehydration, temperature of 102 Fahrenheit, and lung crackles bilaterally with an oxygen saturation of 95% on room air. The attending physician wants to give an injectable antibiotic and asks the patient's mother specifically for a history of drug allergies. The mother denies any, including the specific mention of the antibiotic to be given. After receiving this medication, the child develops urticaria, worsening wheezing, and drooling. Oxygen saturation falls, and the child requires intubation. After giving appropriate treatment and stabilizing the patient, the physician prepares to explain these events to the anxious mother.

This case is an example of a mistake that was unavoidable. It was a mistake to have given this child the antibiotic, causing an iatrogenic anaphylactic reaction. However, there was no way of knowing in advance that this child was at risk of developing this serious reaction. The mother will in all likelihood be aware that something has gone terribly wrong. She may have witnessed increased activity around her child, telling expressions on the faces of the healthcare team, and emergent interventions. There will undoubtedly be questions in her mind about the cause of this unanticipated outcome, and her thoughts may range from blaming the medical professionals to blaming herself. The physician must present a clear explanation of the details of the events leading to this outcome and what it means to her child.

Using the above 'Four Tasks of Disclosure' as a guideline, effective communication between this mother and the physician can occur. Before stepping into the room with this anxious parent, the physician may benefit from quickly reviewing these tasks of disclosure in constructing a plan for the interaction. What follows here is an example of such an interaction:

"Mrs. Smith? I am Doctor Jones, we spoke earlier when you arrived. I want to tell you that right now, Johnny is stable. I'm afraid something has happened that we didn't expect. It seems that Johnny has had an allergic reaction. The reaction was serious. We needed to treat him right away before things got worse. I think the reaction was due to the medicine that we gave him. I know that we talked about it before he got the medicine, but sometimes there is no way to predict that this kind of reaction will occur. I think it would be best for Johnny to stay tonight in the hospital. We would then be able to keep a close eye on him and make sure he continues to improve. He is responding well to the treatment we gave him and hopefully he can go home in the next day or so. I am terribly sorry that this happened. It must have been frightening

to see all that has been happening to him. I want you to know that we will keep you up-to-date on everything that happens, and you can ask questions whenever you like. What questions do you have now?"

The above response illustrates the physician giving as much detail as possible at this time, stating what the mother can expect in the near future, and although the mistake was unavoidable, it includes an apology. The apology in this context has a high likelihood of being appreciated as a show of empathy rather than an admission of culpability. The opportunity to ask questions addresses any additional concerns of the mother.

Knowing that the mother in this case may be feeling guilt over giving consent to use the antibiotic will aid the physician in attending to her emotional needs, thus furthering a positive physician-patient relationship.

Scenario 2: Avoidable Error Causes Harm

There is a difference between avoidable and unavoidable errors. In an unavoidable error, it is easier for the physician to deflect blame by explaining the nature of the error — as in the case above. However, for an avoidable error, the physician's candor will reveal that there was a human factor that could have prevented the bad outcome, and this opens the door for patients and families to blame members in the healthcare team. While providing details about the error, physicians should not speculate about causation at the outset when all the facts may not have been verified. What appears to be the whole truth at the time the error is first discovered may only be a part of the story, and further facts may yet be uncovered. If the physician speculates about the cause, the family may be angrier when the story changes after new facts are discovered that contradict the physician's earlier explanation. The following scenario illustrates a harmful error that was avoidable.

A 56-year-old gentleman was admitted for pneumonia. He had told the triage nurse on initial presentation that he was allergic to "quinolones." Upon admission, the resident misread the triage note because of poor handwriting and ordered levofloxacin. The patient developed an acute anaphylactic reaction, and required a brief intubation. The attending physician then discovered the error and asked the treating resident to accompany him into the consultation room, where the patient's wife is anxiously waiting.

The natural tendency (other than to hide the error) about this type of error is to name the individual at fault. In this case, the patient had stated his allergy but was erroneously given that type of medication anyway. The physician feels like blaming the resident for misreading the triage note, or the triage nurse for sloppy handwriting. The physician may also worry that the patient's wife will hold him responsible for allowing this error to occur.

The tendency to blame must be identified and diffused before the disclosure interaction with the patient or family. By viewing this error as a system problem rather than an individual problem, the burden is shifted away from any individual healthcare personnel. A computer-assisted informatics system would reduce errors due to handwriting. The printing of allergies on the front of the patient's chart would also decrease the likelihood of a mistake. It is important for the disclosing physician to avoid a blaming or accusatory mindset.

The attending physician should be the one to speak to the family. Including the resident in the disclosure interview shows the openness of the team to accept responsibility by not protecting one of its own. Further, it allows the resident to hear precisely what the patient and family hear so that the information being given by the team is consistent, and it provides an excellent teaching opportunity for the physician-in-training. What follows is an example of how to handle this situation.

> *"Hi Mrs. Johnson. My name is Dr. Ching and this is Dr. Bowers. I'm afraid I don't have good news. As you know, your husband got very sick a little while ago. It seems that he had a serious allergic reaction. We treated him for that, and he is stable now, and doing well. I know that he is allergic to certain antibiotics. It may be that his reaction today is due to the antibiotic he was given. I wish I could tell you that I know how this happened, but I am just not sure yet.*
>
> *We are looking into this, and I want you to know that I am taking this very seriously, as is the hospital. We want to make sure that we learn about what happened so that we can make sure it doesn't*

happen again. What this means right now for your husband is that he will be closely monitored tonight to make sure he continues to get better. I'm so sorry that this happened. I feel terrible about this. It must have been very frightening for you to see everything that happened. I want you to know that we will continue to tell you everything that is going on. Please feel free to ask any questions you may have at any time, day or night. What questions do you have now?"

The above response shows empathy, discloses the nature of the error, and avoids laying blame on individuals prematurely, yet promises to get to the bottom of the incident. By using the words *"I'm sorry"* and *"I feel terrible about this,"* the physician communicates his genuine caring. The patient and family must feel that the healthcare team is open, truthful, and willing to engage in dialogue or answer questions. The patient may feel anger that the error occurred, but is less likely to feel intense direct anger at the physician with this type of caring disclosure. In this way, the physician shows that he is the patient's ally and not an adversary.

Even if the physician should eventually take responsibility for the mistake, this does not necessarily translate into a hopeless malpractice case. For one thing, a lawsuit is less likely to be filed if the patient and family sense compassion and humanity. For another, the plaintiff is still required to produce expert testimony to prove that the conduct fell below the standard of care. Nor does an acknowledgement of a mistake necessarily rise to the level of legal negligence. In a recent case of legal malpractice for example, the court rejected the plaintiff's assertion that because the defendant admitted his mistake, it amounted to malpractice *"as a matter of law."*[65]

Scenario 3: An Error Where There is No Harm Done

One of the most controversial questions is whether we should disclose an error that leads to no harm. Some argue that if no harm has resulted, there is no obligation to disclose the error. Consider the following example:

A 45-year-old woman with no drug allergies is recovering well from routine gallbladder surgery. The surgical resident writes an

---

[65] *Kovacs v. Pritchard*, Santa Clara Cty. Sup. Ct No. CV791479, April 8, 2004.

order for cimetidine. The nurse on duty misreads the order and instead gives clindamycin. Only one dose was given to the patient before she discovers the error. The patient suffers no ill effects.

Under these facts, there may be a strong urge to withhold the information from the patient. Since no harm was done, there is very little likelihood that the patient will ever suspect on her own that something was amiss. One could argue that disclosing such an error would only serve to provide potentially damaging information to the patient, erode trust in the healthcare system, and create suspicion and distrust.

There are two problems with this reasoning. First, the patient's basic autonomy and right to information about her own healthcare is ignored in favor of a paternalistic self-protecting cover-up. Second, if this patient were to discover that such an error had occurred (e.g., during records review for an unrelated purpose), the realization that a cover-up had occurred would be damaging to any potential defense, and would taint the credibility of the healthcare team.

When an error occurs that does not lead to any harm for our patients, we have an excellent opportunity to reinforce the strength of our physician-patient relationship. By displaying apologetic candor in light of an admittedly embarrassing event, the physician can turn adversity into opportunity for enhancing honesty and trustworthiness. The disclosure interview may go something like the following.

> *"Hello Mrs. Jackson. How are you feeling? Everything seems to be going quite well for you with respect to your surgery. I do want to talk with you about something else that has happened. It seems that yesterday, you may have received the wrong medicine. I wanted you to have a medicine for preventing stomach upset. Instead, the medicine you received was a common antibiotic, and there doesn't appear to be any ill effects from getting one dose of that antibiotic, or from not getting the stomach medicine. I am not completely sure how this happened, but I can tell you that both medicines are very commonly used, and the names of these medicines are similar. I am very sorry that this happened. I feel terrible about this. I know this news may be upsetting to hear, but I thought you deserved to know. You are important to me, and I want you to know that I value*

our relationship. I would like to answer any questions you might have now."

By breaking the news of a medication error in this way, the physician is able to communicate genuine caring for the patient. By avoiding a candid discussion, we squander a valuable opportunity to enhance the doctor-patient trust.

Scenario 4: The Near Miss — An Error Discovered Before it Reaches the Patient

We are constantly striving to improve safety within our hospitals and practices. Because of improving redundancy and rechecking within our systems, errors may be discovered before they reach the patient. An example is when the wrong medication is ordered, but the nurse notices the error before the drug is given. This is different from an error that reaches the patient but does no harm. There is no ethical obligation to disclose because the patient was never placed in any actual danger of harm, and we can feel confident that an effective patient safety system has been at work.

*****

The medical contrition approach has caught the attention of *The Wall Street Journal*, which recently featured a front page article on the subject, calling it the doctors' new tool to fight lawsuits. The article narrated the story of a Dr. van Pelt, an anesthesiologist, whose inadvertent injection of a drug caused cardiac arrest in his patient, Linda Kenney. Her husband *"wanted to kill the anesthesiologist, flatten him."* But the doctor, against hospital advice, wrote a personal letter expressing his deep sadness over her suffering, and followed this up with a direct apology over coffee. Mrs. Kenney, who had planned to sue the doctor, changed her mind. *"I found out he was a real person. He made an effort to seek me out and say he was sorry I suffered."*[66]

It would be overly optimistic to suggest that disclosure of errors will always prevent a lawsuit. A recent study of health plan members' views revealed that patients will probably respond more favorably to physicians who fully

---

[66] *The Wall Street Journal*, May 18, 2004, pp. A-1.

disclose than those who are less forthright. However, the authors cautioned that the specifics of the case and the severity of the outcome also affect patients' responses, and *"in some circumstances, the desire to seek legal advice may not diminish despite full disclosure."*[67]

Still, disclosure is preferable to silence or a cover-up. It is the right thing to do.

---

[67] Mazor, KM et al. Health Plan Members' Views about Disclosure of Medical Errors. *Ann Int Med* 2004; 140:409–18.

SUMMARY POINTS

MEDICAL ERRORS AND THEIR DISCLOSURE

- To err is human, and medical errors are better viewed as the result of a systems failure than any individual's fault.

- Being competent and careful will not eliminate all medical errors.

- Disclosure of errors to patients and their family demonstrates physician candor, enforces the doctor-patient trust relationship, and results in fewer lawsuits.

- Choose words carefully when disclosing errors in order to show honesty and caring. It is best to avoid finger-pointing or premature admission of fault.

- Patients want to know details of what happened and how the error has affected their health. They value an apology and the promise of truthfulness.

22

Understanding Malpractice Insurance

Everyone recognizes that we are now in another cycle of soaring medical malpractice insurance premiums. Although availability is not the issue despite the withdrawal of several large companies from the marketplace, e.g., St. Paul's, the rate increases have been a focal point of provider protest and political activism reaching all the way to the federal government, the Congress and even the White House. Earlier chapters have reviewed the contentious reasons for the dramatic increases in claims frequency and severity, and insurance rates. This chapter addresses what the physician ought to know, in practical terms, about his or her malpractice coverage.

Incidentally, the malpractice phenomenon is not unique to the U.S. or to the medical profession. The issue of escalating premiums is evident elsewhere. In England, for example, annual subscriptions were reportedly just £2 in 1962. By the late 1990s, GPs were paying £2,000. The legal profession, likewise, faces an increasing number of lawsuits for attorney malpractice. Like doctors, most practicing attorneys are covered by professional liability insurance, although the premiums are substantially lower. For a half-million dollar claims-made policy, one company in 2001 charged a first-year annual premium of $1,500 for a solo legal practitioner. In high-risk areas of legal practice such as securities, estate planning, and wills, the cost rises by 50% or more.[68]

Understanding malpractice insurance means understanding coverage inclusions and exclusions, limits of liability, nose and tail coverage, and the services and strengths of the insurer, be it a commercial carrier or a physician mutual. If one is contemplating going bare, i.e., go uninsured, one should

---

[68]Attorneys Liability Protection Society (ALPS) Policyholder Application 2001.

clearly understand the wide and risky consequences of such a move. It is foolish to think that one can successfully shield all of one's assets and become judgment-proof.

COST

One of the strongest professional liability insurers, the MIEC group (Medical Insurance Exchange of California), urges potential insureds to look beyond cost of coverage. Its brochure urges the doctor to *"look at the potential insurer's financial strength and security, longevity in the marketplace, service to policyholders, claims handling and specific coverage provisions and exclusions."*[69] Still, cost is one of the first things on a doctor's mind when it comes to malpractice insurance. Premiums vary greatly from state to state, and even among locales within a given state. Premiums are of course specialty-dependent, with the highest in the high risk specialties such as obstetrics and neurosurgery. The data in Table 11 shows premiums for the state of Hawaii, a moderate malpractice state. The figures are adapted from MIEC's published 2003–4 Claims-Made Premium Schedule, which classifies doctors into 14 categories according to risk and increasing premiums.[70]

These figures are lower than what is widely believed. However, it should be emphasized that these are first-year rates. According to MIEC, claim-made rates follow a stairstep pattern, rising for five years until a mature level is reached. Furthermore, Hawaii is not a crisis state. The figures cover first-year premiums for a claims-made, not occurrence based, $1 million/$3 million policy. Note that the lowest category, the allergist or psychiatrist (Class 1), pays an annual premium of $1928, whereas neurosurgeons belong in the highest risk group, Class 14, and they pay $26,288. The general practitioner who performs no surgery pays $1928; if he or she delivers babies, the premium rises sharply to $12,620, the category of the general surgeon.

---

[69]What To Look For: A Buyer's Guide for Physicians by MIEC Group. The company is rated "A" by A. M. Best. MIEC is the West's first physician-owned professional liability insurance company. Its address is 6250, Claremont Avenue, Oakland, CA 94618. Tel. 1-800-227-4527. Website is at www.miec.com.
[70]The categories shown have been simplified somewhat. Each category may feature exceptions and variables that may affect premiums, e.g., FPs doing deliveries, Orthopods doing spinal surgery, etc. Obtain full and up-to-date details directly from the company.

Table 11: Claims-made Annual Premiums (MIEC Data, Hawaii, 2003–4)[71]

Class 1:	Allergy, Pediatrics *behavioral*, Psychiatry *including child psychiatry*	$1,928
Class 2:	General Preventive Medicine, Public Health	$2,456
Class 3:	Pathology, Physical Medicine and Rehabilitation	$3,156
Class 4:	Cardiology, Dermatology, Endocrinology, Family/General Practice *no surgery*, Gastroenterology, Hematology, Hospitalist, Infectious Disease, Internal Medicine, Nephrology, Nuclear Medicine, Occupational Medicine (*not industrial*), Oncology, Pain Management Pediatrics, Pulmonary Diseases, Rheumatology, Urgent Care Medicine	$3,508
Class 5:	Industrial Medicine, Ophthalmology	$4,032
Class 6:	Neonatology, Ophthalmology *including refractive surgery or 5% or more cosmetic surgery*, Radiology *including radiation oncology*	$4,384
Class 7:	Assisting at Surgery; Cardiology *including cardiac catheterization and angioplasty*, Dermatology *including hair transplants and liposuction*, Family/General Practice *less than 5% of practice from surgery*, Internal Medicine *including cardiac catheterization and angioplasty*, Neurology	$5,260
Class 8:	Anesthesiology, Urology	$7,012
Class 9:	Emergency Medicine, Family/General Practice *5% or more of practice from surgery*, Otolaryngology	$7,888
Class 10:	Gynecology	$8,764
Class 11:	Colon and Rectal Surgery, Family/General Practice *including obstetrics*, General Surgery, Hand Surgery, Head and Neck Surgery, Orthopedics, Otolaryngology *5% or more cosmetic surgery*, Pediatric Surgery, Plastic Surgery, Thoracic Surgery *excluding cardiovascular surgery*	$12,620
Class 12:	Cardiovascular Surgery, Orthopedics *including spinal surgery and use of chymopapain*	$17,524
Class 13:	Obstetrics and Gynecology	$21,032
Class 14:	Neurological Surgery	$26,288

[71] Taken from MIEC application brochure for 2003–4.

The premiums charged are directly related to both the likelihood and severity of the risk of a claim. For example, a noninvasive cardiologist is at lower risk, and the premium is accordingly lower, similar to that of the general internist at $3,508. On the other hand, once invasive procedures are undertaken, the cardiologist's premium rises to $5,260. The frequency of claims against anesthesiologists has fallen dramatically in recent years because of improved monitoring programs. However, the severity of claims is characteristically very high. Thus, they have been classified as Class 8 (together with the urologist) and pay an annual rate of $7,012. Discounts are generally available to the doctor who is new in practice, as well as for part-timers who practice less than 20 hours a week.

Substantially higher insurance rates are evident elsewhere, especially in crisis states (see Chapter 1: Situation Critical, Prognosis Guarded). Figures tend to be skewed to the high side whenever metropolitan data are used to define the state average. In addition, the numbers do not always represent annual premiums for a $1 million/$3 million claims-made policy. In some states, the limits may be lower, e.g., $200,000/$600,000, and doctors pay an additional surcharge for excess coverage. Indiana, Kansas, Louisiana, Nebraska, New Mexico, Pennsylvania, South Carolina, and Wisconsin are examples of such states. On the average, internists in metropolitan areas pay an annual premium of $12,000 for a $1 million/$3 million claims-made policy.[72]

CLAIMS-MADE VERSUS OCCURRENCE POLICIES

In reviewing insurance rates, one should keep in mind that they may pertain only to so-called 'claims-made' policies, i.e., coverage provided only for claims that are filed for incidents that both occur and are reported while the insurance policy is in force. In contrast, an 'occurrence' policy is one where the doctor is covered for all malpractice incidents during the time the policy was in force, irrespective of whether the doctor is still insured with that particular carrier when the lawsuit is brought. Virtually all policies are now claims-made.

An example: Consider a doctor who purchased and maintained the usual claims-made malpractice policy for the years 1970–1998.

---

[72] Rice, B. Malpractice Premiums: Soaring Again. *Medical Economics,* December 9, 2002, pp. 51–2; *Medical Economics,* January 9, 2004; *Medical Liability Monitor,* 2003 Rate Survey.

He retired in 1999, and stopped his insurance. In the year 2000, a former patient filed a malpractice suit against him for an injury that occurred in 1998. Because he was no longer insured with the company in the year 2000, the year the lawsuit was filed, his claim is not covered. If suit had been brought in 1998, he would have been protected. If the policy were an occurrence rather than a claims-made policy, he would retain coverage for the 2000 lawsuit.

Naturally, occurrence policies are more expensive than claims-made policies, as a suit may be brought many years after the initial negligent act, and the doctor may have retired, died, or switched insurance carriers. This lag period between the expiration of the insurance policy and the bringing of suit is termed the 'tail' ('nose' coverage, a similar concept, is for prior acts coverage when applying to a new insurer). Thus, the difference between claims-made and occurrence policies is the coverage for suits brought after the healthcare professional is no longer insured by a particular insurance carrier, even though the event at issue occurred when the policy was in force. In other words, Occurrence Policy = Claims-Made Policy + Tail.

Virtually all malpractice insurance policies today are claims-made, and typically a doctor who retires or switches companies will have to purchase tail coverage to maintain protection. In the example given above, if the doctor had purchased a tail policy upon his retirement, he would be able to spend his retirement years in repose — free from concerns over malpractice losses.

WHAT IS COVERED?

Malpractice insurance policies typically pay for all legal fees including attorney fees, expert fees, and discovery and court costs. They pay up to the limit of the policy if negligence is found and damages are assessed, either in court or in a negotiated settlement. The latter is the outcome in over 90% of all malpractice claims. Coverage may also extend to liability arising out of peer review activities and other hospital and educational activities, and for the negligence of one's employees. Because these areas of potential liability are not uncommon, it behooves doctors to carefully review the scope of their policy as some carriers do not cover these liabilities. Coverage is usually not provided for intentional torts, e.g., assault and battery, or criminal activities, e.g., Medicare fraud.

A $1 million/$3 million claims-made policy means the limit for each claim is $1 million, and the limit for all claims in a given policy year is $3 million. If the judgment exceeds the policy limits, the doctor is personally liable for the remainder. This has caused fear among some doctors as their personal assets may then be at risk. A reassuring recent article in *Medical Economics*[73] put it this way: *"In theory yes. But in reality, doctors rarely lose their personal assets."* Reasons why both sides usually settle for the policy limit or less include: (1) Until and unless they settle, the plaintiff would receive nothing for medical expenses, and expert and court fees. Typically these fees run into the tens of thousands of dollars which have already been paid (usually fronted by the plaintiff attorney, to be reimbursed by the plaintiff); (2) The lawyer's contingency fee may well amount to nothing unless there is a settlement; (3) The defense may appeal the decision to a higher court, especially when the damages are large, and this can delay payment by years, or even wipe out the judgment entirely if there is a reversal of the verdict; (4) Fear of backlash against the trial lawyers for publicity surrounding a doctor's lost assets; (5) There are usually other deep pockets, e.g., the hospital, to go after in the same case to jointly reach or approach the award amount. Caveat: Lawyers have been known to go after an uninsured doctor or one who is especially arrogant in order to 'punish' them.

Most policies allow the doctor to make the final decision regarding whether to settle and for how much. In other words, the doctor must give his or her consent before any claim is resolved. Look for the language in the 'consent to settle' clause of the policy. However, some carriers retain the final decision-making authority. Occasionally, a doctor refuses to settle for an amount within the insurance limits, preferring instead to proceed to trial. Should the doctor lose at trial, and the judgment is in excess of the earlier settlement amount, he or she may be personally liable for the difference, even if the amount is still within the policy limit. Some policies provide for this eventuality by containing such a provision, popularly termed 'the hammer' as a way of persuading the physician to settle.

MALPRACTICE INSURERS

Doctors are indemnified by a variety of carriers including commercial insurers that specialize in malpractice liability. There are pros and cons to a

---

[73] Rice, B. Could a Malpractice Mega-Verdict Wipe You Out? *Medical Economics*, May 23, 2003, pp. 89–91.

commercial carrier. They are usually large professional business organizations with strong reserves, but they are profit-oriented, and may leave the insurance market if the situation proves unprofitable. Many doctors today are covered by physician mutuals, also called bedpan mutuals, which are non-profit and generally unwilling to settle frivolous suits or insure doctors with a bad track record. Such risk retention groups were authorized by federal law in 1986 to underwrite malpractice liability. By 2003, there were just under 20 such retention groups across the U.S. Not all risk retention groups are successful. The Tennessee-based Doctor's Insurance Reciprocal went into bankruptcy in 2003, leaving some 3000 doctors scrambling for coverage.

MIEC, a physician mutual headquartered in Oakland, California, has been a steady major carrier in several states including Hawaii.[74] Another large carrier, The Doctor's Company, is a doctor-owned company with nationwide reach, and it insures the faculty at the John A. Burns School of Medicine in Hawaii. It prides itself in resolving 80% of claims without indemnity payments or reports to the National Practitioner Data Bank, and of winning 80% of suits that go to trial.[75] In addition, a well organized self-insured group of local doctors called the Hawaii Association of Physician Indemnity (HAPI) has been in existence since 1977. HAPI was formed *"by doctors for the sole purpose of protecting each other from medical malpractice suits and to avoid unnecessary costs that are usually associated with insurance carriers."* Their rates are competitively low, and they insure a significant and growing number of physicians in the state.[76]

CHECKLISTS

The AMA has issued guidelines to its members on how to assess potential insurance carriers, and how to evaluate an insurance policy.[77] The key points to consider about insurers are:

1. Financial Stability
2. Protection against Insurer's Insolvency

---

[74]Website is at www.miec.com.
[75]Website is at www.thedoctors.com.
[76]Website is at www.hapihawaii.com.
[77]AMA: Assessing Potential Insurance Carriers, and Evaluating an Insurance Policy. Available at their website www.ama-assn.org/ama/pub/category/4545 and www.ama-assn.org/ama/pub/category/4584. Accessed June 5, 2004.

3. Performance Record
4. Handling Claims

As for evaluating an insurance policy, the AMA identifies these 14 issues:

1. Type of coverage
2. Limits
3. Premiums
4. Deductibles
5. Definition of a claim
6. Claims reporting
7. Tail coverage
8. Cancellation
9. Extent of coverage
10. Policy exclusions
11. Practice variations
12. Defense costs
13. Consent to settle
14. Additional coverages

Below is reproduced a checklist on professional liability insurance modified from a 1988 publication.[78] This checklist is directed at commercial insurers that underwrite malpractice liability policies:

When choosing the insurance company, ask:

1. Is the carrier financially strong, and what is its rating by A.M. Best, Inc.?
2. If the coverage is offered by a profession-owned 'captive' or a risk retention group, what safeguards does it offer against insolvency? Can insureds be required to make additional contributions?
3. To what degree is the insurance company regulated?
4. Is the company experienced in the professional liability area?
5. What is the quality of defense counsel retained to represent the company's insureds?

---

[78] Council of Insurance, *JADA* 1998; **117**:779. Copyright © 1988 American Dental Association. Reprinted by permission of ADA Publishing, a Division of ADA Business Enterprises, Inc.

When evaluating the scope of coverage, ask:

1. Does the policy provide coverage for all treatments and procedures performed in your practice?
2. Does the policy cover your professional corporation or partnership as a separate entity?
3. Is protection afforded for the acts or omission of your employees?
4. Does the policy protect you against actions arising from service you provide on professional committees, such as peer review panels?
5. Does the policy pay for defense costs during an administrative hearing which could lead to professional discipline or restrictions on one's practice?
6. Do you have a choice of defense counsel?

When considering coverage limits, ask:

1. What is the maximum amount for any single claim?
2. What is the total amount of protection (the aggregate amount) for all claims reported in a single policy year?
3. Are the above limits reduced by the costs of defending your claim?
4. Does the policy contain a deductible provision?
5. Does the policy provide that no claim can be settled without your consent? If so, does your refusal require you to pay an out-of-pocket share of the award should you lose in court?

When purchasing a claims-made policy, ask:

1. Is the carrier contractually obligated to offer an extended reporting endorsement (tail coverage)?
2. Does the tail coverage offer an unlimited length of time to report claims?
3. Is the tail coverage provided without cost in the event of disability, death, or retirement?

SUMMARY POINTS

UNDERSTANDING MALPRACTICE INSURANCE

- All practitioners should understand the extent and limits of their malpractice insurance coverage.

- In selecting a carrier, look beyond cost and consider stability of the insurer, the scope of coverage, and the availability of tail coverage.

- Make sure your insurance policy is paid up and kept current.

- Avoid going 'bare.'

23

What to Expect Now That You've Been Sued[79]

According to the American Medical Association, approximately one of every six practicing physicians in the United States faces a medical liability claim each year.[80] This chapter will explain in simple terms the anatomy of a medical malpractice lawsuit, including tips for a successful outcome. It is divided into ten brief sections as shown in Table 12:

Table 12: Anatomy of a Lawsuit

1. Recognizing the Warning Signs
2. The Summons and Complaint
3. Choosing an Attorney
4. Responding to the Lawsuit
5. Building your Defense
6. The Written Interrogatory
7. The Oral Deposition
8. Settlement Negotiations
9. The Trial
10. Getting on With Your Life

RECOGNIZING THE WARNING SIGNS

Oftentimes the first clue that a doctor might be facing a lawsuit comes directly from the patient. It is not usually as obvious as the patient yelling

---

[79] Dr. Sheri Meslinsky, Pediatrics resident in Buffalo, assisted in the research for this chapter during a senior medical student elective.
[80] Maves, MD. Disappearing doctors: The Medical Liability Crisis. American Medical Association. Available at http://medicine.osu.edu/alumni/disappearingdocs.ppt. Accessed November 23, 2005.

Table 13: Warning Signs

- The *dissatisfied* patient or family
- Those who openly *complain about other physicians*
- *Internet-savvy* patients who research their condition extensively
- *Angry patients*, whether or not the root cause of the anger is medical
- Anyone with a *poor doctor-patient relationship*
- Patient with a *history of filing lawsuits*

this intention, but suspicious patient behavior may include questioning the care provided, asking for records, or changing doctors. Some patients are more likely to sue than others, and Table 13 summarizes some of their characteristics.[81]

The good news is that the vast majority of patients will probably never file a lawsuit. As a physician, the best thing to do is to establish good rapport with patients and their families, keep up to date with medical advances, and double-check the medical records to make sure that they are as accurate and comprehensive as possible.

THE SUMMONS AND COMPLAINT

The first indication of an impending malpractice lawsuit is a request for medical records from a patient or an attorney.[82] Requesting and reviewing the medical records constitute an initial assessment that any diligent attorney is obligated to make before filing a complaint. In some states such as California, a letter of intent to sue must be sent out 90 days in advance.[83] In some 20 states, a malpractice claim is initially sent to a state agency responsible for screening out frivolous claims. The physician would first receive a more benign-appearing letter from such an agency rather than formidable-looking

---

[81] Belli, MM and Carlova, J. *For Your Malpractice Defense.* Oradell, New Jersey: Medical Economics Company Inc., 1986.
[82] Gorney, M. Coping with the Bad News — Part I. *The Doctor's Advocate.* Available at http://www.thedoctors.com. Accessed January 16, 2004.
[83] Nichols, JD. Lawyer's Advice on Physician Conduct with Malpractice Cases. *Clin Orthop* 2003; **1**:14–8.

court papers.[84] If a doctor should perceive *any* indication of a possible malpractice suit, he or she should promptly notify the malpractice insurance carrier.[85]

Receipt of a 'Summons and Complaint' signifies that one is *definitely* involved in a lawsuit. The purpose of the Summons is to inform the defendant that a legal action has been started. The Complaint is a listing of the allegations made against the doctor. One must respond to the Complaint within a designated time, after securing the help of defense counsel.[86]

CHOOSING AN ATTORNEY

One should choose a lawyer as carefully as a patient chooses a doctor. The insurer will appoint one from a pool of experienced attorneys. Most insurance attorneys are able, and some are even outstanding. However, an occasional one may be barely passable. Table 14 lists reasons to seek a new lawyer.

Rarely, the defendant may need more than one attorney. An example is a lawsuit in which one or more claims may not be covered by insurance.

Table 14: You Need a New Lawyer if Yours ...

- has little or no experience in medical malpractice defense
- appears poorly prepared
- does not know the facts of the case well
- does not quickly grasp key medical concepts
- has a bad reputation in the community
- has little or no experience in the courtroom
- does not seem to be 'on your side'
- is not trustworthy
- is impossible to get along with

---

[84] Culley Jr., CA and Spisak, LJ. So You're being Sued: Do's and Don'ts for the Defendant. *Cleve Clin J Med* 2002; 69:752–60.
[85] Ritter, MA and Ritter, NN. A Malpractice Episode: A Sequence of Events. *Clin Ortho* 2003; 1:25–7.
[86] Fish, RM, Ehrhardt, ME and Fish, B. *Malpractice: Managing Your Defense*. Oradell, New Jersey: Medical Economics Company Inc, 1985.

Another is where there are disagreements with the insurer over the disposition of the case. The defense attorney is supposed to represent the doctor who is the true client, although the insurer pays the attorney's bill. In theory, the attorney must represent the doctor to the exclusion of other interests, including the insurance company. When conflicts cannot be readily resolved, retaining a separate attorney may be necessary to ensure separate and independent representation of one's interests and to determine if there are improprieties or a perception of impropriety, both of which represent serious breach of legal professional responsibility.

Once the doctor has secured a capable attorney, he or she should not do anything without the attorney's knowledge and consent, including discussing the lawsuit with anyone else. The mode of communication between the doctor and the attorney is also important. Whenever possible, it is a good idea to speak to the lawyer in person. Remember that the doctor will be sharing with the attorney possible shortcomings and although such communication is confidential and protected from discovery, written information is always potentially dangerous. Be sure to keep any notes, etc., in a file that is *separate* from the patient's medical records.

RESPONDING TO THE LAWSUIT

The initial response to the Summons and Complaint is called the 'Answer' and it is written by the defense attorney. The Answer defines the grounds for the defense, admits or denies allegations, and lists arguments for preventing recovery of damages. The Answer may include a counterclaim, which, however, may not be advisable in medical malpractice cases. An example of a counterclaim would be to sue for payment of an outstanding medical bill. Also included in the Answer may be an impleader, which is an action against a third party. An impleader is included if the defendant believes that someone else is wholly or partially liable for the initial action being brought.

While the lawyer is busy preparing the Answer, the doctor can help by securing his or her own records pertaining to the patient-plaintiff. During the process of Discovery, the defense attorney will be served with 'Notices to Produce' (see below). These are demands for medical records, X-rays, CT scans, or any other documents or files. It is prudent to have secured such files from the moment one is made aware of the lawsuit, so that they can be produced without delay. A significant delay in production or a failure to produce some or all requested files can be damaging to the doctor's case.[87]

BUILDING YOUR DEFENSE

As the attorney builds the defense in preparation for trial, it is important for the defendant doctor to play an active role. For starters, the doctor should know the case better than anyone else. While carefully reviewing the patient's chart, the doctor may notice that something is missing. Do not add to or alter the records except under very special circumstances, and only with the approval of your attorney. It is tempting to want to correct or add on an item or two in the records. However, changes can be detected by modern forensic methods, and they will always place the physician in a bad light even if the intent is innocent, e.g., to clarify or complete the record. As one scours the plaintiff-patient's medical records, it may be useful to create a succinct summary. This can also help your attorney understand some of the medical intricacies involved. A summary is particularly useful in refreshing one's memory before trial, as several years may have passed. Make sure any such summary has "For My Attorney/Confidential" written at the top of it so there is no question it is confidential and privileged information.

A very important aspect of the malpractice defense is choosing an expert medical witness (see Chapter 7: Expert Testimony). Experts are needed to establish the standard of care and to testify that one has exercised the appropriate clinical skills. The expert may be in the same field, or may be a specialist who is willing and able to speak authoritatively. More often than not, multiple experts are retained. Experts must be willing and able to testify at trial, as this is more effective than a taped deposition. It is not recommended to use a friend, partner, or business associate. Most experienced attorneys will avoid the use of professional experts or 'hired guns.'

DISCOVERY: THE WRITTEN INTERROGATORY

The process of 'Discovery' allows the parties to do investigative research and thoroughly review all of the evidence in order to strengthen their

---

[87] Discussion of Medical Malpractice. Bruce G. Clark and Associates. Available at http://medicalmallaw.wld.com/discussion.html. Accessed March 10, 2004.

Table 15: Discovery Devices

Device	Description
1. Requests for Admissions	Parties are asked to admit or deny certain stated facts, and what is admitted need no longer be proven at trial.
2. Notices to Produce	Documents served to the defense attorney demanding any documents, files, X-rays, etc., pertaining to the plaintiff.
3. Demand for a Bill of Particulars	A request for further information regarding the complaint; served to the plaintiff's attorney from the defense.
4. Written Interrogatory	A series of questions in writing that requires a written response under oath.
5. Oral Depositions (Pretrial Testimony)	An oral interrogation of both parties and key witnesses that occurs under oath and recorded verbatim by a court reporter.

respective sides of the case in anticipation of trial. Table 15 lists the various Discovery devices.

The Written Interrogatory is a very important Discovery device used by the plaintiff's attorney to further develop the facts or legal and medical foundations of the case. It is a series of questions about the case that are served to the defense, the answers to which are given under oath and are admissible in court. Responding to the interrogatory is a trying and tedious task, as it may be a 50-page document filled with repetitious and exasperating questions. Of course, the defense attorney will assist the doctor in this task. The defendant must be professional and accurate with the answers as these may be read to the jury during trial, and jurors are not likely to side with an angry, sarcastic doctor. It is equally important to be consistent and to avoid giving contradictory or false information.

DISCOVERY: THE ORAL DEPOSITION

The Oral Deposition is the most important part of Discovery. The individuals called upon to be deposed are called the deponents. The deponents typically include the plaintiff, the defendant, and all of the key witnesses. Only one person is deposed at any given time, and one is only required to be present at one's own deposition. However, the defense attorney may advise sitting in on other depositions as well. The deponent is first questioned, under oath, by the deposing attorney, who may be the plaintiff's or defendant's attorney. This is followed by questions by the opposing attorney. A verbatim stenographic record by a court reporter is made of the entire session, which may last several hours or more. The purpose of the deposition is for each party to learn as much as possible about the opposition's case, thus theoretically eliminating any surprises in the courtroom. The deposition also serves as a means for each party to attempt to gather material that may prove damaging.

An event as important as the deposition deserves a good deal of preparation, which must be done in collaboration with one's attorney. Table 16 lists some pointers on what to do and what not to do.

Table 16: Deposition Tips

- Meet with the attorney for a pre-deposition conference
- Know the current standard of care regarding the case
- Meticulously review patient's medical records
- Bring all requested information (approved by your attorney)
- Be completely truthful
- Give short, concise responses
- Pause before answering question
- Refer to medical records to refresh your memory
- Volunteer no information
- Portray confidence with humility without appearing arrogant
- Read the typed deposition *thoroughly* before signing it
- Find out if your deposition is going to be videotaped. If so, dress appropriately.

Do not be fooled by the relative informality of the deposition, as it may in effect be the defendant's day in court. Most malpractice cases are resolved on the basis of evidence and events presented in the deposition. If the defendant is an excellent witness, the plaintiff's attorney may try to resolve the case short of a trial. On the other hand, if he or she is a very poor witness, the plaintiff attorney will want the doctor to appear before a jury. Even if the case is strongly in the doctor's favor, the defense attorney may recommend an out-of-court settlement if he or she does not think the doctor can stand up well in court.

SETTLEMENT NEGOTIATIONS

A settlement is usually an agreement to pay some amount of money to the plaintiff, without admission of liability, guilt, or wrongdoing. The terms of the settlement can vary, so long as they are agreed upon by both parties. Most medical malpractice cases are eventually settled, but the settlement rarely occurs before discovery is completed. Usually the assigned trial judge will hold a conference with the attorneys in an attempt to arrive at a settlement. If none can be agreed upon, the trial ensues.

Settlement eliminates the defendant's risk of losing, but does not necessarily constitute a favorable outcome to the case. Oftentimes, the defendant-physician is left without an adequate sense of closure. In addition, all malpractice settlements are reported to the National Practitioner Data Bank (a central federal repository of medical malpractice settlements and judgments, as well as reportable adverse actions taken against physicians), so the named physician no longer has a clean slate.[88]

Client consent is usually, but not always, required before the attorney can settle the case. The contractual terms of the insurance policy dictate whether the insured has the right to refuse. The first question to ponder is whether or not malpractice has been committed; one needs to be absolutely sure. Try to be honest. Losing at trial may mean paying a greater amount of damages, not to mention the loss of time and wages, and the adverse publicity. Another important consideration is when one is being sued for an amount

---

[88]Satiani, B. The National Practitioner Data Bank: Structure and Function. *J Am Coll Surg* 2004; 199:981–6.

in excess of the policy limit, which then puts the doctor's personal assets at risk.

Even if the doctor believes that he or she had exercised reasonable care and met the appropriate standard of care, the case could still be decided at trial in favor of the plaintiff, although approximately 75% of trials end up with verdicts for the physician. Injured patients are compelling witnesses, and the jury will have the natural urge to help them. Proof beyond reasonable doubt is not necessary since medical malpractice is a civil, not a criminal matter. The level of proof is preponderance of the evidence, i.e., more likely than not. The American jury system has been much maligned especially in medical malpractice cases, but it is not without its supporters.[89]

THE TRIAL

After months, even years of preparation, the case may finally be brought to trial for judgment before a jury. The trial will start off with opening statements, allowing the attorneys to summarize what they anticipate proving. Next, the plaintiff's case will be presented. This is when the plaintiff and his or her witnesses will take the stand. The defendant's case is presented next. The defendant doctor will be called as a witness, followed by defense experts and any other key witnesses. The defendant will undergo direct questioning, termed direct examination, by his or her own attorney and then a cross-examination by the plaintiff's attorney. The trial is summarized in the closing arguments, as the attorneys recap the important evidence and testimony and argue why their side should win. After the closing arguments, the jury will weigh the evidence and arrive at a verdict.

The importance of the doctor's own testimony cannot be emphasized enough, and thus the doctor should spend ample time preparing it with his or her attorney. The time on the stand is a great opportunity to influence the jury. If the doctor is likeable and seemingly honest, the jury will be less likely to find liability. Jurors appreciate a kindly, gentle physician-defendant, and dislike one who is arrogant and defensive. The table below offers some tips on what to do and what not to do while on the stand.

---

[89]Vidmar, N. *Medical Malpractice and the American Jury.* Ann Arbor, The University of Michigan Press, 1997.

Table 17: Your Time on the Stand

DO's	DON'T's
• Show up on time	• Appear frazzled or hurried
• Dress nicely, but conservatively	• Dress in expensive, flashy clothes
• Respect the judge	• Try to be too chummy with the judge
• Know your case well	• Refer to the medical records too often
• Relax and trust your attorney during direct examination	• Relax and trust the plaintiff's attorney during cross examination
• Give answers consistent with those in your deposition	• Lie or fabricate information
• Speak confidently and in lay language	• Appear arrogant or condescending
• Show compassion for the plaintiff-patient's injuries	• Show guilt for the plaintiff-patient's injuries
• Maintain your composure	• Become hostile, sarcastic, or rude

GETTING ON WITH YOUR LIFE

Being named in a malpractice lawsuit elicits a range of emotions, none of them very pleasant. Some physicians respond with anger and frustration, others with fear and anxiety. Often the legal action is taken as a personal blow, resulting in a loss of self-esteem. Physicians may become depressed and their personal and professional lives consequently suffer. Some have turned to alcohol or other substance abuse in order to help themselves cope with the stress. The doctor will naturally take the lawsuit very seriously, but should *not* take it too personally. It may be helpful to consider a medical malpractice lawsuit as nothing more than an occupational hazard. Remember, to err is human.[90]

---

[90] Gorney, M. Coping with the Bad News, Part II. *The Doctor's Advocate*. Available at http://www.thedoctors.com. Accessed January 16, 2004; Brazeau, CMLR. Coping With the Stress of Being Sued. *Family Practice Management* 2001; 8:41–4.

Finally, and most importantly, one must get on with one's life and practice. Dwelling on the lawsuit after it is over does not serve any purpose. After the trial, one should confidently go back to one's profession, caring for patients. Remember the Hippocratic Oath: *"May I always act so as to preserve the finest traditions of my calling and may I long experience the joy of healing those who seek my help."* The joy of healing does not end because of one 'lousy' lawsuit.

SUMMARY POINTS

WHAT TO EXPECT NOW THAT YOU'VE BEEN SUED

- After receiving notice of an impending lawsuit, promptly contact one's insurance carrier.

- Choose a competent defense attorney, and seek his or her advice on all matters related to the case.

- Master the facts of the case and the relevant literature.

- A professional demeanor, a confident posture, and a humble tone are winning ways at depositions and at trial.

SECTION III

REFORMING THE SYSTEM

24

Medical and Legal Reforms

Alarmed by the sharp rise in medical malpractice lawsuits, a wide coalition of interested parties including government, insurers and healthcare professionals, have looked for solutions or alternatives to our current fault-based legal system. These efforts are usually referred to as tort reforms. However, if the prime obligation is to reduce adverse patient outcomes, prevention being better than cure, then genuine efforts to improve physician training, competence, communication, and the overall efficiency of healthcare system should deserve top attention.

For true reform to occur, we must take an honest first look at our own house of medicine. The overriding goal is to protect patient safety and promote better overall quality of care. This chapter will begin with medical reforms that can lead to the reduction of preventable medical accidents. It ends with a review of tort reforms, including no-fault and modified no-fault proposals.

MEDICAL REFORMS

Medical School

Reducing patient injuries begins with a competent healthcare provider, so it starts with the way we screen and admit students into medical school. Whether admissions committees pay sufficient attention to altruism is always at issue. They can require evidence of 'servanthood' by looking to see if an applicant has performed community volunteer work, especially in health-related fields. Another criterion needs to be compassion. Yet a third is communication and interpersonal skills. Clearly, grades alone are inadequate as a gauge of what the final product will look like. Straight-A students are no more likely to end up as exemplary doctors than a well-rounded B student with qualities of altruism, empathy, and humility.

Academic medicine continues to re-invent ways to shape the curriculum at both the undergraduate and graduate levels, to ensure physicians are well trained to serve patients. One area of growing emphasis in medical schools is the inculcation of the zest for life-long learning, since the half-life of medical knowledge is short. The idea is to instill in student-doctors the curiosity of discovery, encouraging them to be seekers of knowledge, not rote learners, or copiers of lecture notes. The problem-based learning model that is being adopted at a number of medical schools is a step in that direction. The encouragement of research by medical students, mandatory in some schools such as Yale, is yet another. Finally, there is increasing emphasis on the teaching of ethics and communication skills in today's medical schools.

Students are trainees in the truest sense of the word, so their close supervision on the wards is of the highest priority. Senior medical students may be allowed to write orders, including those for medications, under direct and careful supervision. All orders are countersigned before they can be carried out.

Residency Programs

Postgraduate programs are also under constant scrutiny and revision, and ethics, health-law, ambulatory skills and end-of-life care are recent mandated curricular changes. Board certification exams remain the final hurdle to cross, but re-certification at periodic intervals has emerged as a way of encouraging the practicing doctor to keep up with new knowledge and clinical skills.

Specialty boards are charged with the responsibility of reviewing training programs for compliance with guidelines and regulations promulgated by the respective specialty organization. Some of these requirements touch on curricular content, such as elective objectives, percentage of time in ambulatory settings, in-training exams, etc. Others address conditions of training, such as caps on the number of admissions, amount of continuous time on-call and supervision by senior residents and faculty. The overarching principle is reconciling the goals of providing service to patients with learning in a structured and systematic fashion. Audits of training programs occur at periodic intervals, usually every three years, and those that do not meet specified standards risk losing accreditation.

Continuing Medical Education (CME)

Medicine constantly changes and keeping up is part of our pledge. We should not allow a busy practice to break this promise of self-renewal. For the doctor who has completed training and is now in practice, the world of CME purports to provide a convenient way to keep up. Many locales have instituted mandatory CME credit hours as a requirement for license renewal. Unfortunately, the outcomes data are less than assuring, as practice patterns do not change materially after most lecture-style CME programs, and other approaches such as using the case-method or small-group workshops, may yet prove to be better alternatives. Those who need the learning most are also apt to be the ones to thwart the system. No CME committee or enforcement agency can prevent the recalcitrant doctor from dreaming away at a conference.

The arrival of the informatics age should help. Habitual usage of PubMed or Up-To-Date for the latest in diagnosis, pathogenesis and treatment should sustain the passion for learning. Medicine becomes interesting and exciting again, and this translates into optimal patient care.

The Work Environment

This must be conducive to good medical practice. The support staff (nurses, technicians, clerks, etc.) should be well trained and patient-friendly, the facilities uplifting, and the time allotted to each patient encounter realistic, so that nobody feels rushed, and the patient is properly evaluated and educated. Serious attention to feedback from physicians, staff and patients will assist in producing a safe, effective and pleasant work place. An all-too-common situation is the overbooked clinic, where errors and patient harm are just waiting to happen. Far too many doctors are attempting to meet oppressive work schedules. For most patient encounters to be meaningful and effective, 10 minutes for a follow-up appointment, and 20–30 minutes for a new consultation in internal medicine are a minimum.

"Sharpen the Saw": To be excellent, one needs to be refreshed, to recharge. In his book: *The 7 Habits of Highly Effective People*, Stephen Covey calls this *"sharpening the saw."* He tells the story of the woodcutter sawing furiously without stopping to rest or to sharpen the saw that had obviously gone blunt.

As a result, he was far less efficient and effective in his task.[1] We, too, must sharpen our saw as we feverishly pursue our profession and are caught up in the heavy demands of clinical duties, much like the woodcutter and his logs. To maintain excellence, we need to strike a balance between work, education and rest, with time out for family and hobbies.

Medical Errors: The Success Story in Anesthesiology[2]

According to the Institute of Medicine, medical errors cause nearly 100,000 hospital deaths annually. Many of these errors are believed to be preventable. Preventing or minimizing errors in the course of medical diagnosis and treatment can go a long way in reducing adverse events and therefore malpractice claims.

Anesthesiologists as a group have done a magnificent job in reducing anesthetic mishaps and ensuring patient safety. Among other things, their research has led to the use of computerized mannequins that allow rehearsals in simulated life-threatening scenarios, monitoring devices for oxygen, carbon dioxide and carbon monoxide levels, and alarm systems. As a result, anesthesia-related deaths have declined dramatically from one in 5,000 cases two decades ago to one in 250,000. Malpractice premiums have decreased correspondingly. This year, the annual average premium is $20,572 compared with $32,620 (inflation-adjusted) in 1985, a decrease of 37% over 20 years. Regrettably, however, the rates have actually been climbing since 2002 — by about 24%.

LEGAL REFORMS

Justice, compensation, and deterrence are the ostensible objectives of the present tort system. Advocates of this system assert that its objectives are equally well accomplished in medical torts as they are in other personal injury situations. They contend that negligent conduct can be ascertained, and that fault-based tort actions effectively deter substandard care by the medical profession that has failed to police itself.

---

[1] Covey, S. *The 7 Habits of Highly Effective People.* Simon & Shuster, New York, 1989.
[2] Hallinan, JT. Once Seen as Risky, One Group of Doctors Change Its Ways. *The Wall Street Journal,* June 21, 2005, pp. A-1.

Medical and Legal Reforms   265

Others point to the many problems involved in the use of the tort system for medical injuries. They argue that the present system leads to unfair and unreasonable results. Only the few who file an action stand a chance of being compensated for their injuries and even then, they typically wait many years. It is said that the system returns some 28 cents out of every insurance dollar to the injured. The rest is simply gobbled up by the system — legal, expert, and court fees, and other transactional overhead. Furthermore, critics cite the tort system as the cause for the recurrent cycles of widespread insurance liability crises, of which medical professional liability is but one example.

(1) Just compensation for avoidable injuries;
(2) Deterrence of bad medical practice;
(3) Maintenance of insurance affordability and reduction of transaction costs; and
(4) Preservation of the doctor-patient trust relationship.[3]

The adversarial nature of the tort system is not well suited to settling doctor-patient disputes. Because it may be impossible to evaluate and assess fault in many instances, physicians commonly believe themselves to be unjustifiably exposed to adverse publicity, victims of hindsight. Capitulation to unreasonable patient demands may result.

The rest of this chapter will describe the various tort reforms that seem to be under perpetual discussion, as well as review no-fault and modified no-fault proposals. Particular attention is drawn to the Alternative Medical Liability Act (AMLA), a clever proposal first advanced by Professor Jeffrey O'Connell some 30 years ago.[4]

Tort Reforms

In the 1980s, the Attorney General's Tort Policy Working Group released its findings in strong support of tort reform, and concluded that *"while there are a number of factors underlying the insurance availability/affordability crisis,*

---

[3]Tan, SY. The Medical Malpractice Crisis: Will No-Fault Cure the Disease? *U Haw Law Rev* 1987; 9:241–74.
[4]O'Connell, J. No-Fault Insurance for Injuries Arising from Medical Treatment: A Proposal for Elective Coverage. *Emory L J* 1975; 24:21.

tort law is a major cause."[5] The Group recommended the following eight reforms to combat the crisis:

(1) Retain the fault-based standard
(2) Base causation on credible evidence
(3) Eliminate joint and several liability
(4) Limit non-economic damages
(5) Provide for periodic payments
(6) Reduce verdict amount by payments from collateral sources
(7) Limit contingency fees
(8) Encourage alternative dispute resolution

The American Medical Association has taken the lead in the current crusade. A recent poll showed that 72% of Americans want some form of congressional action on tort reform.[6] Not surprisingly, the powerful American Trial Lawyers Association is opposed to tort reform.[7] No cohesive federal package has yet emerged, although H.R. 4280 of the Health Act of 2004 recently passed the U.S. House of Representatives by a vote of 229 to 197. The legislation would limit non-economic damages to $250,000, allocate damages in proportion to degree of fault, and shorten the time interval between injury and the filing of a lawsuit.

Cap on Non-economic Losses: Caps on non-economic damages are the most popular of all tort reforms, and represent a key feature of the current congressional effort towards reforming the system. This proposal limits the amount recoverable for non-economic losses such as pain and suffering, but allows for full recovery of economic damages. The rationale is to provide some predictability to the amount of damages that can be recovered. Non-economic damages are difficult to quantify, and jury sympathy may result in unrealistically high payments. In 1975, the California legislature enacted a $250,000 cap on non-economic loss, and this statute has withstood constitutional challenge. It is believed that California's cap reduced the median

---

[5] Office of the Attorney General, United States Department of Justice. Report of the Tort Policy Study Group on the Causes, Extent & Policy Implications of the Current Crisis in Insurance Availability and Affordability, 1986.
[6] AMA. New Poll: Americans Demand Liability Reform. Available at www.ama.assn.org. Accessed June 6, 2004.
[7] Association of Trial Lawyers of America, whose website is at atla.com.

malpractice award by $366,000. The figure is especially significant in neonatal injuries where reductions were imposed in 71% of cases, with a median value of $1.5 million. As more and more states enact caps on non-economic losses, some lawyers are reportedly turning away cases where economic damages, which are not capped, are not large enough to warrant taking the case to trial.[8]

Still, not everyone agrees that the system needs fixing. It has been pointed out, for example, that *"Overall, the total payout for medical malpractice insurers in the United States is about $4 billion a year, which is about half of what we spend annually on cat and dog food."*[9]

The plaintiffs' bar opposes limiting both economic and non-economic damages, arguing that such limits unfairly discriminate against the victim with severe injuries, such as the quadriplegic or the brain-damaged neonate.

A variation on 'caps' is the creation of state operated patient compensation funds, or malpractice insurance pools that are financed by surcharges on physicians or general revenues. The idea is to assist physicians by taking over payments in excess of their liability limits. South Carolina, for example, will pay out of such a fund any malpractice damages over $200,000 per incident.[10] Unless prudently financed, such funds may face insolvency, as happened in Hawaii.

Regulation of Attorney Fees: This is the most controversial tort reform measure, with doctors and lawyers on opposite poles of the debate. America is one of very few developed nations in the world that allows a contingency fee system. So entrenched is this mode of legal conduct that no group has seriously proposed its outright ban. Instead, the usual proposal is to regulate attorney fees, seeking to reduce the amounts plaintiff attorneys would net in a malpractice case. The idea is that this reform will return to the victim a larger portion of the damages, or reduce the actual amount awarded.

Is it reasonable for a plaintiff attorney to regularly take a third or more of the victim's award? If so, a $3 million award for a claimant wrongfully injured

---

[8]*The Wall Street Journal*, October 8, 2004, p. A1, citing Rand Institute for Civil Justice as source of the figures.
[9]Doroshow, J. Are Caps the Answer to the Malpractice Crisis? *Internal Medicine News*, December 1, 2004, p. 8.
[10]South Carolina Code §38-79-420.

would net the lawyer $1 million or more in fees. This windfall is a direct function of the seriousness of the injuries, not the complexity of the case. Hence the paradox: the severity of the injuries, not the egregious conduct of the tortfeasor, determines whether the plaintiff's lawyer takes the case in the first place.

The Tort Policy Working Group has recommended the following sliding scale and contingency fee schedule: 25% for the first $100,000; 20% for the next $100,000; 15% for the next $100,000; and 10% for the remainder. Thus, for a jury award or settlement of $1 million, the plaintiff's attorney would receive $130,000 rather than $333,333, assuming a one-third contingency fee.

Trial lawyers argue that the outright removal of the contingency system will rob plaintiffs of litigating a meritorious claim, and will deprive penniless victims of their 'keys to the courthouse.' It is extremely expensive and time-consuming to mount a malpractice lawsuit — and always a gamble for the plaintiff attorney who must front the expenses, typically in the tens of thousands of dollars. They lose many of their cases, and therefore depend on 'windfalls' to absorb the risk of loss of fees and costs, not to mention the 2–5 years it takes to bring a case to closure. Regulation or abolition of contingency fees interferes with the free market and may make it unprofitable to sue. The net result arguably would be to jeopardize the victim's chances of selecting and obtaining competent counsel.

Abrogation of Joint and Several Liability: Under the doctrine of joint and several liability, every defendant who is determined to be at fault for the plaintiff's injury may be held responsible for the <u>entire</u> damage award, irrespective of the degree of fault. This '1% law' means that a 1% negligent defendant is potentially liable for all damages in the event other defendants cannot be joined or are unable to pay their proportionate share. Opponents of joint and several liability argue that it is unfair for a defendant to pay more than his or her proportionate share. They assert that the rule encourages plaintiffs to seek a deep pocket defendant, which is invariably the doctor or hospital. This explains why hospitals insist that physicians carry adequate liability insurance before they are granted medical staff privileges. The Tort Policy Working Group has recommended the abrogation of this doctrine, except in the limited circumstance where the plaintiff can demonstrate that the defendants have actually acted in concert.

Supporters of joint and several liability contend that it is better for the tortfeasor, irrespective of degree of fault, to fully compensate the innocent victim, than for the victim to be under-compensated. They emphasize that the minimally negligent tortfeasor is no less a 'but-for' cause of the injury.

Some states, e.g., Kansas, have enacted legislation to apportion the amount of damages according to the degree of fault of the defendant.[11] Others have used a threshold fault percentage above which the defendant is required to pay the entire damages. In Pennsylvania, that figure is 60%.[12]

A modification of joint and several liability is enterprise liability, which shifts liability from physician to the enterprise, e.g., health plan(s). Thus, the enterprise will bear all responsibility for damages. This idea was advanced in the Clinton health reform proposal of 1993, but has not proven popular.

Abolition of the Collateral Source Rule: Under this rule, if an injured person receives compensation from a source wholly independent of the tortfeasor, then the payment will not be deducted from the damages. By making evidence of collateral payments inadmissible, this law creates the situation where the injured victim may in fact be doubly compensated. Abolition of the rule would reduce the damages by an amount equal to collateral payments derived from health and disability insurance and other sources. Supporters of the collateral source rule contend that a victim should not be penalized for prudence in buying insurance protection. Additionally, subrogation rights already require the victim to reimburse health insurers in the event of tort recovery.

To support dual insurance coverage for the same injury is to endorse an inefficient system with duplicative premiums. Collateral payments should be subtracted from jury awards to prevent double dipping, because subrogation is not a significant consideration in many tort actions, especially medical malpractice cases. Some commentators have proposed abolishing subrogation to avoid duplicative premium payments and to lower transaction costs.

Some states such as Florida now require awards to be reduced by payments from collateral sources, less any insurance premiums already paid for by the plaintiff.[13]

---

[11] Kansas Statutes §60-258a(d).
[12] 42 Pa. Cons. Stat. §7102(b.1).
[13] Florida Statutes §768.76.

Other Tort Reforms: Other reform proposals include:

1. Mandatory structured payments in which periodic rather than lump-sum payments are made;
2. Penalties for the filing of frivolous suits;
3. Modified statutes of limitations, reducing the time interval during which suit can be brought for injuries to infants and minors;
4. Stricter standards for expert witnesses;
5. Affidavits of non-involvement where the physician can file an affidavit denying involvement, and the plaintiff must present evidence at a hearing that establishes a reasonable basis for instituting an action against the individual doctor;
6. Making the 'loser' pay all attorney fees and court costs;
7. Arbitration or mediation;
8. Screening panels.

Arbitration or Mediation: Mandatory or optional arbitration or mediation are sensible suggestions, as these methods of dispute resolution are less combative, more efficient, and cheaper and faster in bringing closure. Arbitration requires an appointed qualified arbitrator, who sits like a judge, directing the proceedings which include the admission of evidence. The arbitrator's decision cannot usually be appealed, although they can be vacated on rare occasions under strict rules. Mediation, on the other hand, looks to the parties in dispute to come to a solution that is acceptable to both sides, and the mediator's job is to facilitate the reaching of this mutually acceptable position rather than to be judgmental.

These methods have worked well in some instances. Although there are an insufficient number of arbitrators or mediators who are specifically trained to look at medical issues, these professionals are generally much better qualified than the average juror. Unfortunately, relatively few doctor-patient encounters incorporate a contract to submit disputes to arbitration, and these usually relate to disputes over fees rather than to injuries.

Notwithstanding these limitations, the Kaiser HMO system of healthcare has successfully utilized mandatory arbitration for virtually all its malpractice claims. And the judiciary in Hawaii has mandated binding arbitration for all tort claims of less than $150,000 under its Court-Annexed Arbitration Program, although either party may reject the arbitrator's award.[14]

---

[14] Hawaii Arbitration Rules, Rule 6.

Medical and Legal Reforms   271

Screening Panels: Occasionally, the plaintiff files a lawsuit that is without merit, or names everyone who is even remotely connected with the alleged negligence. Many jurisdictions have therefore set up mandatory pre-trial screening panels with the objective of weeding out frivolous or nuisance suits. However, some of these panels have been found to be unconstitutional. In 1976, Hawaii created the Medical Claims Conciliation Panel (MCCP), which remains popular among Hawaii's doctors.[15] All medical negligence claims are reviewed by this screening panel before suit can be filed. The panel consists of three members, i.e., two lawyers and a doctor, and strict rules of evidence are not enforced during the hearing. The decision of the panel, favoring the doctor in about 70% of the time, is non-binding, but the process is mandatory before a lawsuit can subsequently be filed. The idea behind the MCCP is to weed out frivolous claims at an early stage. However, because the findings are non-binding, the parties can use the proceedings as a fishing expedition to determine how strong a case the other side has. Critics contend that the MCCP hearing merely prolongs the litigation process, and increases costs without substantial corresponding benefit. In Ohio, pre-trial screening panels were tried and abandoned for these reasons.

California's Medical Injury Compensation Reform Act (MICRA)[16]

Many of the above reforms were passed by the California legislature in 1975 under the Medical Injury Compensation Reform Act (MICRA). In the preceding ten-year period, the number of claims in California had tripled, and claim severity had increased 1000%. The preamble to the Act identified the legislature's concern over the malpractice crisis:

> *"The Legislature finds and declares that there is a major health care crisis in the State of California attributable to skyrocketing malpractice premium costs and resulting in a potential breakdown of the health delivery system . . ."*

MICRA limits non-economic recovery in medical negligence cases to $250,000 and permits juries to be informed of collateral source payments. Attorney contingency fees are placed on a sliding scale, being limited to 10% for damages over $200,000. Periodic payments for future damages are written into the law, and plaintiffs are required to give a 90-day advance notice of

---

[15] Hawaii Revised Statutes §671-5(b).
[16] Medical Injury Compensation Reform Act of 1975, Cal. Civ. Proc. Code § 3333.2 (West 1982).

an impending claim to encourage settlement negotiations before proceeding to trial. The California Supreme Court has ruled that these reforms are constitutional, since they are rationally related to the legitimate legislative goal of reducing medical costs.

The positive effects of MICRA are palpable. Shortly after MICRA was enacted, California hospitals received a rebate of their premiums. A study commissioned by the California Medical Association concluded that MICRA had been effective in holding down claim costs, which reduced the annual rate of increase from 15% in the pre-MICRA years to 7% post-MICRA. The cost of professional liability insurance appeared to have stabilized as a result of MICRA. At one point after MICRA was passed, the California obstetrician paid an average of $40,000 annually compared with $94,000 in New York, and $154,000 in Florida. From 1983 to 1989, the average increase in premiums was 97% for California obstetricians, while the national average increase was 248%.[17]

Many tort reformists hail MICRA as the prototype success-story, crediting it for bringing California's premiums from one of the highest in the nation, second only to New York City, to one of the lowest over the past 25 years. Eleven other states have since implemented similar reforms.[18] By a 51–49 margin, voters in Texas recently approved Proposition 12, which limits non-economic damages to $750,000, with $250,000 from the doctor and $500,000 from the hospital. The amount cannot be appealed to the courts. Preliminary data suggest that the cap is working. Texas Medical Liability Trust, the state's largest malpractice carrier, has reduced rates by 17% under the new law.[19] In addition, Proposition 12 allows the legislature to impose caps on other types of lawsuits, e.g., punitive damages in products liability cases.[20]

MICRA is not without its detractors. A state senator in California is planning to introduce legislation to raise the cap to $900,000, and a consumer group has threatened to eliminate the cap altogether via a referendum vote.[21]

---

[17] Lobe, TE. *Medical Malpractice: A Physician's Guide*. McGraw Hill, 1995.
[18] The Doctors Company: What is MICRA? Available at www.thedoctors.com. Accessed August 22, 2004.
[19] Frieden, J. Texas Doctors Cautiously Optimistic About Tort Reform. *Internal Medicine News*, June 15, 2005, p. 75.
[20] Malpractice Suits Capped at $750,000 in Texas Vote. Available at www.NYTimes.com. Accessed September 15, 2004.
[21] Frieden, J. Doctors Urge California to Keep 'Pain and Suffering' Cap at $250,000. *Internal Medicine News*, November 1, 2004, p. 73.

The No-Fault Solution

Thoughtful academicians such as Professor O'Connell have asserted that the present fault-based tort system is dysfunctional, it is because of the tremendous inefficiency, unfairness and prohibitive costs associated with fault-finding. The injured victim is randomly and unjustly compensated. Additionally, the adversarial nature of the proceedings traumatizes the physician and the doctor-patient relationship, with grave implications for the entire medical care system.

In many instances, fault simply cannot be ascertained when an individual is injured after receiving medical care. Some of these same criticisms have led to the introduction of the no-fault system for auto injuries and for workers' compensation. In general, the abolition of fault-finding has resulted in a more efficient system for resolving auto injuries. The no-fault approach is associated with no increase in negligent conduct, nor has the cost been burdensome. No one would seriously suggest a return to the traditional fault-based system for these injuries.

Not everyone is enamored by the no-fault concept. The Justice Department itself, decrying *"compensation often awarded merely for the sake of compensation,"* has condemned the movement of the tort system towards no-fault liability.

Injuries arising out of medical care differ in one essential aspect from all other injuries — they may be a natural and unavoidable consequence of the underlying illness or treatment. To compensate without regard to fault cannot be taken to mean compensating all adverse medical results. Rather, medical no-fault must embrace in some fashion the concept of compensating avoidable injuries. Herein lies the dilemma — how to identify the avoidable injury, or the so-called compensable event. O'Connell notes:

"We can say with some degree of certainty that an automobile accident occurs because the drivers were driving on the road, and not because of some other pre-existing or extrinsic cause ... In medical care, however, it is far less clear whether the lack of success from medical treatment is the result of improper medical care or merely from the natural workings of disease."[22]

---

[22] O'Connell, J. No-Fault Insurance for Injuries Arising out of Medical Treatment: A Proposal for Elective Coverage. *Emory L J* 1975; 24:21.

If avoidable injuries can be predetermined, a list of compensable events can be drawn up. Professor Havighurst has proposed such an approach under the rubric of medical adversity insurance. While acknowledging the complexities of identifying compensable events and the failure of physician groups to take up the project, Havighurst has, nonetheless, compiled a preliminary list of such events in anesthesiology and general surgery. Examples of compensable events are permanent recurrent laryngeal nerve damage after parathyroid surgery or thyroidectomy, hemolytic reaction following a blood transfusion, and brain injury under anesthesia in patients between six months and sixty years of age undergoing relatively simple operations.[23]

The hope of avoiding case-by-case adjudication of medical injuries and exposure to unmanageable costs is possible only if there is a particularized and comprehensive list of compensable injuries. Presumably, this list would cover avoidable incidents such as laryngeal nerve damage in neck surgery, but such a standard approaches that of negligence, because an avoidable mishap occurs in the absence of due care. Therefore, the end result will be a return to the fault-based tort system.

On the other hand, if all complications of treatment were deemed compensable, including those that are unavoidable, then a true comprehensive no-fault system would exist. It would prove prohibitively expensive. The narrower approach of compensating only preventable injuries is similar to the negligence standard, whereas the broader approach means compensating virtually all injuries. This inherent difficulty in assigning a medical event as compensable has frustrated efforts to fully develop a workable no-fault patient injury compensation system. Preliminary and incomplete lists have been proposed, but further development has not been forthcoming. The more realistic view is that this hurdle will prove insurmountable.

New Zealand's and Sweden's No-Fault Systems: This same dilemma currently faces the New Zealand no-fault compensation system that came into effect in 1974 under its Accident Compensation Act. All accidental injuries including medical injuries are removed from the tort system, and covered by this Act. In

---

[23] Havighurst, C and Tancredi, L. Medical Adversity Insurance — A No-Fault Approach to Medical Malpractice and Quality Assurance. *Milbank Memorial Fund Q.* 1973; 51:125. Reprinted in *Ins L J* 1974; 613:69.

practice, about 40% of malpractice claims are denied. Because of escalating payout costs, a special Medical Misadventure Account was created in 1992 to specifically handle malpractice damages. Professional liability premiums, which are experience-rated, fund this account.[24]

Under the initial scheme, injured patients did not have to prove fault. All the claimant had to establish was *"medical, surgical, dental, or first aid misadventure."* Although negligence was not necessary, what constituted a medical misadventure was not defined. At the same time, the law stipulated that *"not all medical negligence comes within the scope of medical misadventure,"* and common law tort actions for medical negligence remained available. Despite this lack of both definitional and functional clarity, the Accident Compensation Commission was quite definite that it was not necessary to show that there had been negligence on the part of a medical practitioner before a claim for medical misadventure will succeed.

However, the revised Act of 1992 now requires the claimant to show *'medical error,'* which is defined as *"the failure of a registered health professional to observe a standard of care and skill reasonably to be expected in the circumstances."* With this definition, which is the legal language for negligence, the 'no-fault' system of compensating medical injuries in New Zealand has effectively been subsumed by the traditional fault-based tort system.

Sweden set up a no-fault system in 1975 to compensate 'avoidable injuries,' and has the apparent support of Swedish physicians who are more willing to admit errors and assist their patients to apply for compensation if they meet the threshold — 10 days in hospital or 30 days out of work.[25]

The Alternative Medical Liability Act (AMLA)

Some two decades ago, Congress considered the Alternative Medical Liability Act (AMLA) which proposed a clever medical injury compensation

---

[24] Gellhorn, W. Medical Malpractice Litigation (U.S.) — Medical Mishap Compensation (N.Z.). *Cornell L Rev* 1988; 73:170.
[25] Weiss, GG. Malpractice: Can No-Fault Work? *Medical Economics*, June 4, 2004, pp. 66–71.

system that would circumvent the shortcomings of medical no-fault.[26] AMLA incorporated the following key features:

(1) The option of the medical provider to tender payment to the patient for economic loss within six months of injury;
(2) Such tender forecloses future tort action by the injured victim;
(3) Compensation benefits for net economic loss include 100% lost wages, replacement service loss, medical treatment expenses, and reasonable attorney's fees;
(4) Non-economic losses are not reimbursable;
(5) Payment is net of any benefits from collateral sources;
(6) Civil action will not be precluded for compensation tendered for intentional torts and in wrongful death cases;
(7) Compensation benefits would be payable no later than 30 days after submission of proof of economic loss;
(8) Tort action remains available to any claimant absent provider tender of payment.

The appeal of AMLA lies in its provision for prompt, fair and reasonable compensation benefits without the need to inquire whether the event is compensable. It is the healthcare provider who has the exclusive right to tender payment. The financial incentive to do so insures that many more injured patients will be compensated. This incentive derives from the exclusion of non-economic claims, and the offset provided by collateral source payments, and savings from expensive litigation. In short, the proposal rids the present system of inefficiency and randomness, replacing it with efficiency and predictability.

An Example: By way of illustration of how AMLA might work, consider the case of a worker rendered paraplegic following spinal anesthesia. The physician has the incentive to tender payment for all economic losses, including all future wage loss, because such tender will foreclose tort action, with the risk of a far larger award. All parties further benefit by the non-adversarial nature of the transaction, and by the speedy resolution of the dispute. The

---

[26]O'Connell, J. A Neo No-Fault Contract in Lieu of Tort: Pre-Accident Guarantees of Post-Accident Settlement Offers. *Cal L Rev* 1985; 73:898. AMLA was introduced as Congressional Bill 5400 H.R. on June 28, 1984.

patient, in turn, has the security of prompt payment equal to his or her economic loss, without incurring significant attorney or expert fees. In addition, the patient avoids the uncertainty of a long and costly claim, with potentially no recovery. The physician can return to caring for the sick and dying, and the injured patient can concentrate on rebuilding his or her future, confident that the economic needs will be promptly and fully met. Under AMLA, minimal disruption of the doctor-patient relationship should result due to the decrease in litigation.

Although AMLA's trade-off is a smaller compensation package, this is mitigated to the extent that claimants will no longer bear the burdensome price tag of attorney and expert fees. AMLA will hopefully achieve the goal of fairly compensating the largest number of those injured. Allowing the healthcare provider the option of tendering payment further avoids the need to identify compensable events. In essence, the provider is asked to identify whether a given event is compensable or not. The incentive for settling lies in the reduced award and the avoidance of litigation.

Moreover, AMLA can be used as an effective vehicle for quality assurance and peer review. Compensated injuries can trigger an automatic medical review, followed by appropriate educational, rehabilitative, or disciplinary action. Data generated can also be used as the basis for experience rating in liability insurance underwriting. Thus, an effective deterrence against substandard medical practice can be an important byproduct of AMLA. Physicians would not tender compensation for every adverse result, since all such tenders will lead to automatic review for substandard care. Caveat: One should guard against overly harsh and unprincipled disciplinary penalties else one may abort all incentives to tender compensation. This responsibility properly resides in the institutional medical care evaluation committee, which should develop a fair and objective review process to decide the quality of care.

AMLA will also further reduce defense costs by allowing the hospital to adopt 'corporate liability' and to tender payments. So long as any one member of the health provider team tenders payment for the injury at issue, tort action is foreclosed. Thus, by allowing the hospital as well as the physician to tender provides an even greater opportunity to compensate the injured without resorting to the wasteful practice in which, for a single injury, multiple defendants are typically joined as parties, each requiring separate defense counsel. Mechanisms regarding contribution from physician tortfeasor(s) can be easily formulated, with, for example, the use of an objective review panel.

Finally, although estimates of the cost of AMLA have not been made, internalization of costs, with funding through insurance premiums paid by medical providers, appears preferable. Spreading the costs evenly among all patients would be another alternative, but such an approach would destroy the potential use of provider experience-rating to modify premium rates. The adoption of AMLA should not lead to an escalation of insurance premiums, since savings from equitable awards, combined with the abolition of litigation costs and duplicative payments should significantly decrease the cost of insurance.

Despite the many attractions of AMLA, the politics of the time prevented its passage into law. Hearings were held on AMLA, under House Resolution 5400, before the 98th U.S. Congress, 2nd Session, on June 28, 1984. Congressmen Richard Gephardt and Henson Moore sponsored H.R. 5400, which covers recipients of Medicare and other federal programs. Professor Jeffrey O'Connell, its architect, provided testimony in support of the bill. Both the American Medical Association (which favors other tort reforms) and the Association of Trial Lawyers of America opposed the bill. It died in committee, and was reintroduced the following year in the 99th Congress (25 July, 1985) as the Medical Offer and Recovery Act, H.R. 3084. The bill was referred to five committees, but no hearings were held.

Summary of Tort Reforms

Attempts at tort legislation are not expected to substantially improve the medical malpractice situation. These reforms, however, may be useful as temporary remedies, providing a measure of predictability and stability that is needed for continued functioning of the system. The use of arbitration or mediation, as well as screening panels, hold the greatest promise in alleviating the emotional trauma associated with malpractice lawsuits.

A broad no-fault compensation scheme for medical injuries would remove inefficiency, adversariness and unpredictability, but at prohibitive costs. Moreover, the threshold question of what constitutes a compensable event is impossible to define, except in a few narrow areas. Thus, a true no-fault approach to medical injuries is impractical.

A modified no-fault scheme such as AMLA provides a workable solution to correct the inequities of the present system. This proposal invites tender by medical care providers for economic loss in return for foreclosure of tort action. The proposal has the appeal of simplicity and efficiency, and avoids the need of pre-identifying the compensable event, yet achieves the goal of just compensation in a non-adversarial setting. A non-statutory alternative, where the parties contract for this mode of injury compensation, might also be effective. By identifying many more cases of mal-occurrence, AMLA presents the unique opportunity to fashion an effective peer review process and to identify areas for risk management. Regrettably, this bold and innovative idea at genuine tort reform appears to be lost forever to politics.

SUMMARY POINTS

MEDICAL AND LEGAL REFORMS

- Reforming the system means both medical and legal reforms.

- Medical reforms should emphasize life-long active learning, enhance communication and scientific skills, and cultivate attitudes of caring, compassion and humility.

- Curricular rigor, supervision and accountability must characterize all training programs.

- Medical errors are a system problem deserving of close study.

- Some measure of tort reform, e.g., caps on non-economic loss, may be enacted into law to attenuate the insurance crisis, but this will not remove the unfair, random and adversarial nature of the litigation process.

- A modified no-fault system for compensating medical injuries has many merits and may provide a workable solution.

SECTION IV

MULTIPLE CHOICE QUESTIONS AND ANSWERS

For each of the thirty-five (35) multiple choice questions in this section, one or more answers may be correct. Or all may be wrong. I have provided answers with brief explanations (see pp. 300 to 317).

QUESTION 1

Dr. John Sullivan was in the restaurant when a fellow diner choked on his steak. The doctor did not provide first-aid, and the diner subsequently died. A lawsuit against Dr. Sullivan for failure to render emergency aid would succeed because:

A. The doctor owes a duty to treat in a jurisdiction that has the Good Samaritan statute.
B. A reasonable person would come to the aid of someone in distress.
C. A reasonable doctor would come to the aid of a stranger in distress.
D. All doctors have taken a vow to treat in an emergency situation.
E. But for the doctor's negligent failure to treat, the patient would have survived.

QUESTION 2

Unbeknownst to her parents, little Janet, a 3-year-old toddler, accidentally ingested some aspirin pills while mom was away seeing the family doctor for the 'flu.' When she returned, she found that little Janet was sweaty and irritable and was breathing rapidly. She immediately brought the 3-year-old to the doctor, who diagnosed Janet's condition as the 'flu,' similar to her mother's. The doctor prescribed baby aspirin. Twelve hours later, the child died of salicylate poisoning. At trial, the plaintiff's expert testified that the sweating and hyperventilation should have been a tip-off to the doctor that this may have been a case of salicylate poisoning.

A. The family physician is not negligent because he had no way of knowing that there was an accidental ingestion of aspirin.
B. But for the mother's carelessness in leaving the child unattended, the poisoning would not have taken place. Thus, the mother is contributorily negligent.
C. Poisoning is a leading cause of pediatric deaths, and all family physicians are expected to consider this diagnostic possibility.

D. The plaintiff must prove causation to win the lawsuit.
E. The doctor is liable as evidenced by the fact that he prescribed aspirin, which is contraindicated in young children because it can cause Reye's syndrome.

QUESTION 3

A general practitioner (GP) delayed hospital admission for a patient with chest pain who later died from a myocardial infarct (MI). A close friend of the family heard the news over the phone several hours later and reacted with extreme grief. She later lapsed into a prolonged depression that required psychiatric treatment.

A. GP is not liable to the patient because his chances of surviving the massive MI would at best have been improved by 10% had he been hospitalized earlier.
B. As a general practitioner, GP used his best judgment and should not be held to a higher standard.
C. GP is not liable to the family friend as there is no doctor-patient relationship.
D. GP is liable for the friend's injuries because he caused them.
E. Physical injuries are compensable, but psychiatric ones are not.

QUESTION 4

A patient developed severe headache and neck stiffness which the clinic physician diagnosed as a viral infection. Her condition did not improve, so her husband called the doctor who did not return the page. The call was transferred to the emergency department (ED) physician who asked some questions but did not encourage re-evaluation, as the ED was extremely busy at the time. The patient's condition was subsequently diagnosed as a subarachnoid bleed and she later expired. Her husband sued the clinic physician, ED physician, and hospital for malpractice. The clinic physician, a hospital employee, is a medical resident just out of medical school. The ED physician works as an independent contractor and derives no direct salary or fringe benefits from the hospital. A prominent sign at the entrance features these words: "Hospital Emergency Services: Physician on duty 24 hours."

A. The clinic physician is not liable because he met the standard of care expected of a physician at his stage of training.
B. The ED physician is not liable because there was no doctor-patient relationship.
C. The hospital is vicariously liable for clinic physician's conduct but not ED physician's conduct.
D. The hospital cannot be liable because it is not a person.
E. No liability attaches to any of the three parties because their negligence, if any, did not cause the death of the patient.

QUESTION 5

Having lost faith in several of her doctors including her cardiologist, Mrs. Hee decided to seek non-traditional alternative treatment for her heart condition. She received chelation therapy, herbal enemas and megavitamin injections from Dr. Snakeoil, but died after six months of treatment. Dr. Snakeoil is a holistic healer, not a medical doctor, but he is licensed by the State. Mrs. Hee had signed a form stating that she understood the unproven nature of Dr. Snakeoil's treatment methods and she was willing to assume the risks.

Had she taken the advice of her cardiologist, she would have received effective treatment for her cholesterol and blood pressure conditions. Interventional procedures to improve her coronary circulation may have proven beneficial.

In a claim against Dr. Snakeoil, which of the following is (are) true?

A. Mrs. Hee assumed the risk of unconventional treatment, and this shielded Dr. Snakeoil from liability.
B. Assumption of risk is only a partial defense.
C. Dr. Snakeoil had an obligation to refer her to a cardiologist.
D. Mrs. Hee lost her claim because when asked whether conventional treatment would have been life-prolonging, her expert-cardiologist replied: "That would be speculative."
E. Dr. Snakeoil will be judged by what a reasonable holistic healer would have done under the circumstances.

QUESTION 6

During her hospital stay, an elderly patient noticed abrasions and burn marks on her extremities. She believed they resulted from the use of wrist and ankle leather restraints but could not prove it.

A. This is a case of *res ipsa loquitur* or 'the thing speaks for itself,' analogous to the leaving of surgical instruments in the abdomen.
B. If an unexpected adverse event occurs in the hospital, a good case of *res* can be made because the hospital team is in full control of the patient.
C. This is not a case of *res ipsa*, as the injuries may have resulted from excessive rubbing on the bed-sheets.
D. *Res ipsa* is good circumstantial evidence and the plaintiff will no longer need a medical expert to prove her case.
E. The hospital cannot be liable because the leather restraints were necessary to prevent the patient from thrashing around and posing a risk to herself and others.

QUESTION 7

Mary visited her favorite sister Cecilia in the hospital where she had recently undergone brain surgery. During the visit and in full view of Mary, Cecilia developed *status epilepticus* after a nurse gave her Dilaudid instead of Dilantin. Mary was petrified by the incident and developed insomnia, nightmares and depression.

A. Cecilia can sue the nurse for medical malpractice.
B. Mary can sue for the negligent infliction of emotional distress.
C. The hospital may escape liability because the nurse was an independent contractor from an outside agency.
D. The doctor will be sued because his illegible handwriting caused the wrong medication to be administered.
E. The suit will likely fail because seizures are a common post-op event after brain surgery, and the wrongly administered drug may not have been the offending agent.

QUESTION 8

A pathologist came to the aid of a woman who had collapsed in the shopping mall. The woman wore a Medic-Alert bracelet indicating that she suffered from anaphylaxis, and that she carried adrenaline in her purse. The doctor performed CPR but did not administer the drug because he had not given an injection in over 30 years. The patient died and expert testimony indicated that had the adrenaline been given, the patient would have survived.

A. By coming to the aid of the stranger, a doctor-patient relationship was formed.
B. Doctors are ethically bound to treat those who are in need of medical assistance.
C. Doctors are legally bound to treat those who are in need of emergent assistance.
D. No liability attaches since the doctor did not breach the standard of care expected of a pathologist under the circumstances of the case.
E. The 'Good Samaritan Doctrine' covers aid to strangers and generally allows for recovery only if gross, rather than simple, negligence is proven.

QUESTION 9

Doctors most prone to lawsuits:

A. Are busy practitioners.
B. Have poor interpersonal skills.
C. Talk down to patients.
D. Rarely apologize.
E. Belong in high risk specialties such as neurosurgery and obstetrics.

QUESTION 10

The pharmacist filled the wrong prescription, dispensing Diabinese instead of Dilantin. The patient developed hypoglycemic seizures as a result and suffered permanent brain injury. In an action against the pharmacy, which of the following will be helpful to the defense?

A. The doctor's handwriting is illegible and the pharmacist did try to page the doctor, but he did not return the call.
B. The Learned Intermediary doctrine shields the pharmacy from liability.
C. Hypoglycemia is an intrinsic defect of the drug Diabinese and liability should fall on the manufacturer.
D. Strict liability rules govern this type of injury.
E. This was a moonlighting pharmacist not the usual employee of the drug-store.

QUESTION 11

In spotting patients who may be suit-prone, which of the following is (are) true?

A. Patients who are critical of others.
B. Poor patients on welfare.
C. Educated patients who surf the Internet.
D. Doctor-shoppers.
E. Those who have sued before.

QUESTION 12

In order to relieve intractable pain in a terminally ill patient, Doctor D had to administer increasing amounts of morphine. This led to respiratory arrest and hastened the patient's death. Doctor D's action is:

A. A cause in fact but not a legal cause.
B. A proximate cause.
C. The intentional tort of assault and battery.
D. An example of 'double effect.'
E. Homicide.

QUESTION 13

Mrs. Sonnenberg suffered multiple fractures after her car was struck from behind by a drunk driver. The orthopedic surgeon negligently nicked her femoral artery during surgery, which resulted in profuse hemorrhage that required six units of packed red blood cells. Although she survived, Mrs. Sonnenberg was left with irreversible renal failure and she now requires lifelong dialysis.

A. The drunk driver is liable for all injuries that resulted from his negligent driving.
B. The drunk driver is not responsible for her hemorrhagic complications since he did not cause them.
C. The surgeon's negligence was a superseding cause and frees the drunk driver from liability.
D. The surgeon may be sued for malpractice.
E. If Mrs. Sonnenberg did not survive the operation, both the drunk driver and surgeon would be charged with homicide.

QUESTION 14

Dr. E diagnosed a rare endocrine disorder called MEN Type 1, which is inherited in an autosomal dominant fashion. Genetic measurements confirmed the diagnosis, but Dr. E did not offer to inform or counsel the patient's three siblings and two children. Is there any liability in this instance?

A. Dr. E's legal duty was only to his patient, and it was entirely the patient's duty to inform her relatives.
B. He would escape liability because this is a rare and exotic disease.
C. A reasonable doctor would not have offered family counseling.
D. A reasonable endocrinologist would have provided family counseling after securing the patient's permission.
E. Unless injury results, there can be no malpractice action.

QUESTION 15

Although you are a general internist, you regularly obtain and interpret X-rays of your patients instead of having a radiologist read them. What level of accuracy or standard of care will you be held to? Assume that the community standard is for radiologists, not internists, to read X-rays.

A. Other general internists.
B. Board-certified radiologist.

C. Non board-certified radiologist.
D. A standard between a radiologist and a general internist.
E. An X-ray technician whose level of expertise in the field of radiology is similar to yours.

QUESTION 16

Your patient tests positive for HIV but he refuses to disclose this to his wife despite your repeated advice. His wife is also your patient.

A. You should not disclose the diagnosis since the patient has not given you his consent.
B. You have a legal duty to disclose to his wife.
C. You have to disclose to the health authorities as required by law.
D. The patient may sue you if you breach patient confidentiality.
E. The wife may sue you if you keep silent.

QUESTION 17

Regarding iatrogenic injuries that are medication-related, which of the following is (are) true?

A. The doctor is liable if the drug was prescribed for a non-indicated condition.
B. The doctor failed to warn of serious risks.
C. The doctor failed to warn of a rare complication.
D. The patient did not ask about the side effects and therefore was contributorily negligent.
E. Liability will attach to the manufacturer for a "defective product."

QUESTION 18

Dr. DeSouza, overlooking his patient's known allergy to penicillin, erroneously prescribed ampicillin, a broad-spectrum penicillin. Fortunately, his patient Tony discovered the error before taking the medication. On his way back to the doctor's office to obtain a new and different prescription, he was struck by a car and suffered serious injuries.

A. Had it not been for the doctor's negligence, Tony would not have needed to make the return trip. Dr. DeSouza is therefore liable.
B. No medical harm came from the initial wrong prescription. 'No harm, no foul,' and therefore no liability.
C. The accident was most likely a superseding cause, freeing Dr. DeSouza from liability.
D. The accident was most likely a concurring cause, and Dr. DeSouza remains fully liable.
E. The proper party to sue is the careless driver who struck Tony.

QUESTION 19

Following thyroidectomy, the patient developed persistent postoperative hoarseness. The surgery also left an ugly scar. The patient did not ask, and you did not specifically warn, of these complications. She was an anxious patient and you did not wish to upset her unnecessarily. However, she did sign, but did not really read or understand, a consent form witnessed by the nurse that detailed these complications. She now sues you for lack of informed consent, claiming that she would not have undergone the surgery had she known of these risks. Which of the following are valid defenses?

A. She does not have a case because she signed the consent form.
B. Disclosure of risks is not needed, as this is a particularly anxious patient who would be unduly frightened by the information.
C. The medical community standard is not to tell patients about horrible complications unless specifically asked, and you are merely following community practice.
D. Her thyroid condition was a serious one, and surgery was an emergency.
E. She is unreasonable because she is using hindsight in claiming that she wouldn't have undergone the surgery.

QUESTION 20

Through an oversight, the nurse fails to get the patient's signature for the consent form. However, the surgeon did discuss the procedure and risks and did obtain the patient's oral consent. The surgery was complicated by an infection that prolonged the hospital stay.

A. Oral consent is not legally valid, so the surgeon faces a malpractice suit or a claim for the intentional tort of battery.
B. Oral consent is legally valid, and simple backdating can rectify the lack of a signed form.
C. Oral consent is legally valid; the surgeon should write in the progress notes that he had obtained the patient's informed consent.
D. It's the patient's word against the doctor's, and usually the latter will prevail in court.
E. Claims alleging lack of informed consent go hand in hand with claims of substandard care.

QUESTION 21

Giant-size fibroids were discovered during a routine D&C and diagnostic laparoscopy. The gynecologist proceeded with a total hysterectomy because two other colleagues agreed that this was the definitive treatment and the patient was already under general anesthesia.

A. Gynecologist did right as he was thinking of his patient's best interest.
B. Gynecologist did right as there was an implied consent for the hysterectomy.
C. Gynecologist was merely applying the principle of therapeutic privilege.
D. Gynecologist was merely applying the principle of necessity.
E. Gynecologist should have discussed the situation with the patient at a later date and obtain specific informed consent before proceeding with the hysterectomy.

QUESTION 22

Emily recently developed diabetes and was started on insulin therapy. Two weeks later, she lost her coordination while driving and struck the car in front of her, seriously injuring its driver. In the emergency room, her blood glucose was found to be in the hypoglycemic range of 40 mg/dl. Her doctor had not warned her of this side effect of insulin.

A. Emily can sue the doctor for negligence because she was not warned of this risk.

B. Her lawsuit will fail because informed consent is about surgical or other invasive procedures, not about medications.
C. She was herself not hurt, and without physical injuries, there can be no malpractice claim.
D. The driver in the other car can sue her for negligent driving.
E. A lawsuit against the doctor by the driver in the other car will probably fail because there is no doctor-patient relationship.

QUESTION 23

Ms. Holistica purchased Slim-You, a herbal supplement sold over the counter as a weight loss agent. She asked her primary care doctor about its effectiveness and safety, and he said that it was OK. Another patient had used it and lost weight without apparent complications. Two months later Ms. Holistica developed jaundice, abnormal liver function tests, and liver failure.

A. The doctor is not liable because he did not prescribe the supplement.
B. The doctor may be liable because he had given his approval for its use.
C. Ms. Holistica should sue the drugstore for selling an unsafe drug.
D. Ms. Holistica should sue the manufacturer for a defective product.
E. No one is liable unless the plaintiff proves proximate causation.

QUESTION 24

An unconscious man was brought to the Emergency Department in vascular collapse. He had been thrown off his motorcycle and ruptured his spleen. The surgeon recommended emergency surgery but no next-of-kin was readily available to give consent. A card in his wallet indicates that the patient is a Jehovah's Witness, and that he should never receive a blood transfusion.

A. No consent for the operation is necessary as this is an emergency.
B. No consent for the blood transfusion is necessary as this is an emergency.
C. If he desperately needs a blood transfusion, one should be given to him, otherwise he will die.
D. If his spouse can be located, and she gives consent for the transfusion, then it's okay.
E. Operate on the patient, but respect his disavowal of blood.

QUESTION 25

Because of a particularly busy and hectic day, your medical charting was incomplete for a patient whom you saw at 8.00 a.m. You had to leave for a hospital emergency and attend to some pressing personal matters. Your patient had complained of chest pain and you did obtain a normal EKG. You later discovered that he collapsed after he left your office and was hospitalized with a myocardial infarct. It is now 11 p.m., and you have not had dinner. What should you do?

A. Leave the incomplete note as is, and wait until the next day to complete the records when you are less hungry and your mind is fresher.
B. Complete the records before going home.
C. Write 'incomplete' and leave the note as is.
D. Complete the note with an explanation of the interruptions, and include the time of entry of both notes, including the normal EKG.
E. Medical records are kept for the convenience of the doctor. One should spend more time taking care of the patient and less time taking care of the records.

QUESTION 26

A 38-year-old woman first consulted a board certified surgeon for a breast lump. The surgeon found none on examination, reassured the patient, and asked that she return for follow-up in three months. The patient forgot her three-month appointment. A year passed, and she returned complaining of an enlarging mass. The surgeon was now able to palpate the breast lump and promptly scheduled a biopsy, which the hospital pathologist erroneously read as benign.

Eight months later, the patient developed metastatic breast cancer. The surgeon performed a radical mastectomy, but the disease was too widespread to be cured surgically. Had she known that it was not surgically curable, she would not have consented to surgery.

Which of the following statements is (are) correct?

A. The surgeon is incompetent in missing the initial breast lump.
B. The surgeon should have ordered a mammogram.

C. The pathologist is liable because his error caused her 'the loss of a chance.'
D. The patient is contributorily negligent because she failed to keep the return appointment.
E. This is also a case of lack of informed consent.

QUESTION 27

You are treating a patient, a school-bus driver, for cirrhosis and alcoholism. You report both his diagnoses to his employer and he is fired from his job because of the drinking history. He sues you for breach of confidentiality.

A. You are liable for having breached patient confidentiality.
B. You have a normal legal duty to report a patient's medical condition to a legitimate employer.
C. There is a public safety issue here that requires the doctor to breach confidentiality.
D. The patient will probably win the lawsuit if he is an office clerk instead of a school-bus driver.
E. If this is a sick-leave note, you should simply write 'off work for medical reasons.'

QUESTION 28

You operate a for-profit medical website that displays clinical summaries of the latest in diagnosis and treatment. You also answer personal medical questions from subscribing viewers via e-mail. There is a disclaimer that you are not providing medical advice, and visitors to your website are encouraged to consult their own personal physicians. A viewer, in reliance on your information, suffers harm.

A. No doctor-patient relationship is formed in cyberspace, so there is no duty of due care.
B. The disclaimer effectively immunizes you against any lawsuits.
C. Negligence claims are more likely to prevail against for-profit medical websites such as yours than not-for-profit educational ones.
D. You may be liable for breach of privacy or confidentiality if the e-mail messages are intercepted or read by someone else.

E. Answering your patient's questions via e-mail is like a phone consultation, except that everything is documented.

QUESTION 29

Dr. Moon, a general practitioner (GP), saw his patient on two separate occasions for fever and myalgia. She then suddenly experienced the sensation of a curtain covering her left eye. The doctor suspected retinal detachment, and advised her to seek specialist attention, but did not insist on immediate attention as it was Christmas Eve. The patient waited 36 hours without improvement in her vision, and then went to the emergency department, where the rare condition of Klebsiella endophthalmitis was made. She subsequently lost the vision in that eye.

A. In a suit against Dr. Moon, the plaintiff would prevail because Dr. Moon made the wrong diagnosis, and this is malpractice *per se*.
B. Dr. Moon told the patient to see a specialist, and therefore legally discharged his duty.
C. Dr. Moon's failure to obtain a stat eye consult once he suspected retinal detachment is a breach of the standard of care.
D. In order to win the lawsuit, the plaintiff must prove that the 36-hour delay caused her blindness.
E. A plaintiff sees a GP at her own peril, and assumes the risk of GP missing rare or difficult eye diagnoses.

QUESTION 30

A 5-year-old boy develops excruciating testicular pain from torsion of the testis. The urologist recommended immediate emergency surgery but both parents refused. What should the surgeon do?

A. Perform surgery without consent as this is an emergency and consent is unnecessary.
B. If the boy nods in assent to the proposed surgery, this would constitute valid informed consent under the circumstances.
C. Acquiesce to the parents, as they are the rightful custodians of the boy's health.

D. Remind the parents they may be in violation of child abuse statutes if they insist on withholding consent, and proceed to the operating room irrespective of their protest.
E. Request a court hearing, and treat the boy conservatively in the meantime.

QUESTION 31

When signing out to a colleague prior to going on vacation, which of the following is the most important?

A. Someone with board certification in your specialty who works in the same hospital.
B. An emergency room physician who is willing to respond to your patients who may be in need of urgent medical assistance.
C. A colleague with the best attitude regarding covering on a reciprocal basis.
D. A new doctor in town who has lots of time to answer patient calls.
E. It is best to instruct your assistant to refer all calls to the nearest hospital so as to escape liability.

QUESTION 32

For purposes of malpractice risk management, the practitioner is well advised to observe which of the following 'red flags?'

A. An angry patient.
B. An overworked doctor.
C. Careless and aggressive billing practices.
D. Sketchy record keeping.
E. Poor bedside manners.

QUESTION 33

You are about to retire, but it was one of those hectic and unbelievable final days. Your clinic assistant broke a hypodermic needle, which lodged in the patient's deltoid. A patient's son tripped and fell in your waiting room and sustained a fracture. And in a heated and angry meeting you voted with others on the peer review committee of your hospital to suspend a 'rotten apple' doctor. All three victims file suit against you six months into your retirement. What results?

A. Your malpractice insurance will cover all three incidents. That's what insurance is about — it covers professional liability.
B. Only the first incident will be covered, because it's the only one that involves allegations of malpractice.
C. Even the first incident is not covered since your policy covers physician negligence, which does not extend to an employee.
D. You are retired and are no longer insured, so your retirement assets are now at risk.
E. Good planning, you have tail coverage.

QUESTION 34

The newly completed Good Neighbor Hospital was hosting its open house for the neighborhood and community doctors. Mrs. Brown, a 60-year-old woman in previous good health, was enjoying the hospital tour when she suddenly collapsed and became unresponsive.

Mr. Passerby, an engineer who had recently completed a Red Cross cardiopulmonary resuscitation (CPR) course, promptly began vigorous external cardiac compression, managing to crack three ribs in the process. Nurse Nightingale, who works as a part-time dermatology assistant, provided mouth-to-mouth resuscitation, although she appeared to be unfamiliar with the procedure. At about the same time, Dr. Goodnight, a retired general practitioner, came to Mrs. Brown's assistance. He tried unsuccessfully to measure her blood pressure and her pulse and did not administer any drugs.

Six minutes later, the hospital Code-500 team arrived.

A. No one is liable as this is a classic case of immunity under the Good Samaritan statute.
B. Mr. Passerby is the least likely to be found liable as he is a layperson.
C. Nurse Nightingale ought to re-certify in basic CPR.
D. Once a doctor, always a doctor. Dr. Goodnight is negligent.
E. Six minutes is too slow a response time for a Code-500. The hospital is definitely liable.

QUESTION 35

Your office assistant misfiled a critical lab report showing dangerous hyperkalemia. Unaware of the abnormality, you did not notify the renal patient to promptly return for treatment.

A. You are not liable as the patient had no complaints when seen three weeks later despite serum K+ of 6.5.
B. You are liable for the death of the patient who developed ventricular fibrillation.
C. Your office assistant is liable, and so you are liable.
D. The lab is liable as it failed to call you for the abnormal result.
E. Misfiled lab and X-ray reports are a common source of errors leading to patient harm and litigation.

ANSWER TO QUESTION 1: NONE CORRECT

There is no legal duty for anyone, even a doctor, to come to the aid of a stranger, even in an emergency. The law does not use the reasonable person standard to determine duty, only to determine what the standard of care ought to be. The ethical responsibility to assist one in an emergency situation is borne by all doctors, but that does not translate into a legal duty. Hence no suit will prevail, even if the failure to treat causes death. The Good Samaritan statute merely immunizes against a lawsuit arising out of negligent aid, but it does not mandate the giving of aid.

ANSWER TO QUESTION 2: C, D

This question deals with the issue of the failure to diagnose. A common reason for a malpractice suit, the case revolves around the proper standard of care. Pediatricians and family doctors who treat infants should know that the accidental ingestion of medications is a leading cause of morbidity in this patient population. The doctor in this case should have inquired into this possibility, even though the presenting signs and symptoms are also compatible with a viral infection. Since the mother was unaware of the infant's overdose, the history here would be unhelpful or even misleading, so the case will turn on whether the physical findings themselves are enough to raise the suspicion of salicylate poisoning.

The plaintiff's expert testified that an aspirin overdose should have been suspected, suggesting that the doctor was practicing below the standard of care. The defense will have to bring in an expert or more likely several experts to testify otherwise, and the jury will evaluate the credibility and logic of the dueling experts in reaching their decision regarding fault.

It is true that without the mother's own negligence in leaving the child unattended, harm would not have come to the child. Strictly speaking, this is not contributory negligence, which applies to the victim's own negligence, and the victim here is the child. However, a separate third party claim can be entered by the defendant against the mother, and she may also face prosecution for child neglect.

Aspirin is contraindicated in young patients because of the risk of Reye's syndrome, a potentially fatal complication. This obviously places the doctor in a poor light, as it speaks to his overall low quality of care. However, the injury here is death from salicylate poisoning, not Reye's syndrome. Thus the plaintiff must still prove by expert testimony that the doctor's failure to timely diagnose salicylate poisoning was a proximate cause of death. Whether the doctor's prescription of aspirin further aggravated the child's condition and therefore was an additional causative factor depends on whether little Janet was given any of the prescribed aspirin. The facts are silent on this point and she may have been too ill to take any! Choice E is therefore incorrect.

ANSWER TO QUESTION 3: NONE CORRECT

All doctors, including general practitioners, are held to the standard ordinarily exercised by their peers. GPs are not held to the same standard as say, cardiologists, so the plaintiff would have to prove with expert testimony that a GP exercising reasonable care under similar circumstances would have promptly hospitalized the patient. Using one's best judgment may remove the moral culpability but it is not good enough to meet the legal standard. To win, the plaintiff will still have to prove proximate causation, i.e., that the failure to hospitalize was both a factual and legal cause of the patient's demise. Although the risk is reduced by only 10%, many jurisdictions will consider this 'loss of a chance' to constitute sufficient causation.

Answers C, D and E are also incorrect. It is true that there is no doctor-patient relationship between the doctor and the family friend, but this is not required in allegations of negligent infliction of emotional distress. Liability for this tort requires proof of proximity in time and space, and a close relationship to the primary victim. A close relation is usually a family member and physical presence is needed to satisfy the proximity requirement. If these elements are present, there may well be liability even if the injured is a third party with no prior connection to the tortfeasor. The facts in this case do not seem to meet these criteria, so the GP will likely escape liability to the family friend — but not because there was no doctor-patient relationship.

Finally, prevailing case law supports the notion that legitimate psychiatric conditions, such as depression, are compensable without need to show accompanying physical injuries. In the past, emotional injuries were thought to be difficult to define, so the courts insisted that there be simultaneous physical injuries in order for a plaintiff to successfully claim damages.

ANSWER TO QUESTION 4: NONE CORRECT

Whether the clinic physician is liable will depend on whether the original medical history and physical findings were sufficient to raise the diagnosis of a subarachnoid bleed. That he is a medical resident will not in and of itself absolve him of liability. Inexperience is not usually regarded as an adequate defense and most jurisdictions will judge him by the standard expected of a fully qualified doctor.

By talking to the husband and asking questions about the patient, the ED physician had established a doctor-patient relationship. Whether he breached the standard of care by failing to ask the patient to immediately come to the hospital will depend on the questions he asked and the answers he received. Certainly a busy ED is not a good enough reason to keep a patient with a life-threatening condition away.

Any entity, not only a person, can be held liable for civil damages. Hospitals can therefore be asked to pay damages for any number of reasons, e.g., direct negligence, vicarious liability, premise liability, and so on. Although there is still controversy over whether residents are hospital employees, the facts here stipulate an employer-employee relationship, which makes the hospital liable on the basis of *respondeat superior*. That is, the negligence of the servant (employee) is imputed to the master (employer). This is a classic example of what is termed vicarious liability. Since the ED physician is not an employee but an independent contractor, *respondeat superior* does not apply in his case. However, the plaintiff would attempt to cast the ED physician as an ostensible agent of the hospital, and so hold the hospital vicariously liable. As evidence, the plaintiff would point to the hospital sign that the ED doctors are working for the hospital 24-hours a day.

Finally, for the defense to be successful it must counter with expert testimony that the subarachnoid bleed was far too advanced for effective medical intervention, i.e., the disease caused the death, not any delay in diagnosis and treatment. Prompt surgical intervention may be life-saving in a subarachnoid bleed, so the defense faces an uphill fight on the causation issue.

ANSWER TO QUESTION 5: A, D, E

The facts here are clear that Mrs. Hee actively sought the care of Dr. Snakeoil, and had assumed the risk of holistic treatment from a duly licensed practitioner of holistic medicine. Assumption of risk, unlike contributory or comparative negligence, is a complete bar to recovery.

Whether Dr. Snakeoil is obligated to refer the patient to a cardiologist is subject to the standard of care expected of a reasonable holistic healer. This will be established at trial by an expert in holistic medicine and not by an allopathic medical doctor. It is unlikely there is such a duty to refer, especially since the patient in this case had already visited several conventional doctors including a cardiologist.

Finally, the case was lost after the plaintiff's own expert testified that it was speculative to infer causation. In order to prevail on the issue of causation, the expert must state that it was more probable than not that her life would have been prolonged by conventional cardiac therapy — not that it was possible or worse yet, speculative.

ANSWER TO QUESTION 6: C

For *res ipsa* to be applicable, three conditions must be met: (1) the injury would not have occurred in the absence of someone's negligence; (2) the plaintiff was not at fault; and (3) the defendant had total control of the instrumentality that led to the injury. In some jurisdictions, *res ipsa* shifts the burden of proof to the defendant (normally the plaintiff has the burden of proof), but in most jurisdictions, it is merely circumstantial evidence that is rebuttable, so expert testimony will still be needed. In any case, the plaintiff will need an expert to prove causation and damages. The facts here are insufficient to constitute a clear case of *res ipsa.*

Medical mishaps in the hospital occur under complex circumstances, which is why most courts are reluctant to invoke the *res ipsa* doctrine except in very special situations, e.g., sponge left in peritoneal cavity.

Answer E is incorrect. Use of patient restraints, either physical or chemical, is taboo under current JCAHO standards. Evidence indicates that injuries are more apt to occur, not less so, with their use which literally deprives patients of their physical freedom. Other means of treating the patient must be found, e.g., baby-sitters. Pleading this line of defense is therefore unlikely to prevail.

ANSWER TO QUESTION 7: ALL CORRECT

In this case, Cecilia can sue several parties including nurse, doctor and hospital. Mary also has a cause of action against these co-defendants as her emotional injuries occurred in close proximity to her sister's injury. If the nurse is an independent contractor, presumably with separate insurance coverage, the hospital may be dropped from the action unless there is evidence for an ostensible agency or failure to properly credential, supervise, etc. If the doctor's illegible order is shown to be the reason for the error, then the doctor will clearly be at risk. The nurse still has an independent duty to clarify the order especially if it's for an unusual drug or an unusual dose. Finally, as in all malpractice claims, the plaintiff must prove proximate causation by a preponderance of evidence, i.e., the erroneous drug caused the seizures which were foreseeable. The facts here, i.e., post-op brain surgery, suggest that proving causation will be difficult for the plaintiff(s), but this will depend to some degree on the medication's propensity for inducing seizures.

ANSWER TO QUESTION 8: B, D, E

This case scenario involves a doctor treating a stranger, so there is no formation of the traditional doctor-patient relationship. The law does not require one to come to the aid of strangers. This extends to medical practitioners, although there is an ethical but not a legal duty to do so in the case of an emergency. Thus, choices A and C are incorrect. To encourage aid to strangers, many jurisdictions have enacted statutes to immunize aid-givers against being sued if negligence on their part results in harm. However, if gross negligence is proven, there may still be liability.

The pathologist's failure to administer adrenaline does not amount to gross negligence which usually means reckless disregard of the risks and consequences.

ANSWER TO QUESTION 9: ALL CORRECT

The first four choices speak to poor attitude and communication skills. These predictably get doctors into trouble. The last option is also correct. Here the reason is the catastrophic and tragic injuries arising out of negligent care — often at birth and in patients with complicating brain damage.

ANSWER TO QUESTION 10: A, E

The pharmacist will be judged by the reasonably prudent pharmacist standard. His attempt at paging the doctor may or may not have been enough to satisfy this standard. The plaintiff will surely argue that he should have held off on filling the prescription, attempt to reach the doctor at a later time or try to reach his alternate.

The 'Learned Intermediary' Doctrine is not applicable here. It holds a doctor liable for adverse reactions from the use of medications and shields drug manufacturers from liability.

The hypoglycemic action of Diabinese may be an intrinsic property of the drug but the case here deals with the wrong prescription, not a side effect arising out of a medically indicated prescribed and dispensed drug. The law of strict liability applies to ultra-dangerous activities or defective and unreasonably dangerous products, and is unlikely to apply to Diabinese-induced hypoglycemia.

The drugstore's best defense is the fact that the errant pharmacist was moonlighting and functioned as an independent contractor, not as a store employee. In other words, the drugstore is hoping there will be no finding of vicarious liability.

ANSWER TO QUESTION 11: A, C, D, E

The majority of patients do not sue their healthcare providers even though there may have been a negligent act leading to injuries. Demanding and well-educated patients are more likely to sue, as are those who are already familiar with the legal system and with lawsuits. Watch for the hyper-critical patient. Financial gain is not the usual reason for a malpractice lawsuit. Poor and uneducated patients are <u>less</u> likely to sue as they may not have the medical sophistication to recognize substandard treatment and they usually lack the know-how to seek legal redress.

ANSWER TO QUESTION 12: D

The 'double effect' phenomenon describes the situation where a foreseeable adverse outcome supervenes even though the intent is to provide a beneficial effect. The principle is clinically invoked to permit the aggressive use of comfort measures such as pain-killers in terminally ill patients — even though they may hasten death. It is cognizable in both law and ethics as legitimate and acceptable practice.

The usual rationalization is to find that the act did not cause the death of the patient (no causation; the underlying terminal disease, not the narcotic, caused the death), or to find no duty to treat where the withholding or withdrawal of life-sustaining treatment is at issue. The very rare prosecution of physicians for homicide under these types of circumstances has never been successful.

Even if we accept there is causation, a civil suit would fail because there is no breach of the standard of care. This is also not an assault and battery, which is an intentional act that causes apprehension of or actual offensive touching without consent. Consent in this clinical setting has usually been explicitly given by the patient or the surrogate decision-maker, or is implied.

ANSWER TO QUESTION 13: A, D

The surgeon's negligence would most likely be construed as a foreseeable event, and therefore it constituted a concurring, not a superseding cause. This makes the drunk driver liable for all injuries including the bleeding and renal failure. Option A is therefore correct.

D is also correct. The surgeon himself may be sued for malpractice and may be found liable for the hemorrhagic complications if the nicking of the artery is shown to be a negligent act. This is by no means a foregone conclusion, as the measure of negligence is what is to be ordinarily expected of a surgeon under the circumstances. There may have been extenuating circumstances such as an obscured surgical field that led to the mishap.

The drunk driver would certainly be charged with homicide if Mrs. Sonnenberg died, as there was gross negligence or reckless disregard (drunk driving). The surgeon, on the other hand, would not be so charged as there was no intention on his part to cause death, and the surgical mishap was at most simple negligence.

ANSWER TO QUESTION 14: D, E

The law requires that one acts reasonably, and for a doctor this means the standard expected of one who has the knowledge, skill and judgment ordinarily possessed by fellow members of the profession. The standard of care in this case would probably require that an endocrinologist offer counseling to a patient with a genetic condition. This would include advising the patient to bring her family members in for screening or to arrange for comparable follow-up with another physician. A non-specialist may or may not be expected to adhere to this standard. The former is likely, as family counseling is regularly emphasized during medical training. Expert testimony would be required to determine whether this is the standard for non-endocrinologists. At the minimum, there is a separate duty to refer to a specialist if the condition falls outside of the treating doctor's ability. The test is not whether the disease is rare and exotic, but what a reasonably prudent doctor would do under the same or similar circumstances.

For a malpractice suit to succeed, the plaintiff must prove causation and damages. If no injuries can be traced to the negligent act, there will be no case. Failure to offer genetic counseling, however, may well lead to subsequent injuries to affected individuals.

ANSWER TO QUESTION 15: B, C

Every doctor is held to the standard of his specialty. However, if one assumes the duties of another specialty, the law will consider you as holding yourself out as one who is capable of functioning at that higher level. If internists do not regularly read their own X-rays and you, an internist, choose to do so, you will be held to the standard of a radiologist. The standard expected of a radiologist, however, is not dependent on board-certification.

ANSWERS TO QUESTION 16: B, C, D, E

Only upon the basis of trust can a patient begin to form a relationship with a doctor. This means the doctor must respect the patient's confidentiality, so without consent, medical information cannot be disclosed.

However, under some circumstances a doctor is obligated to breach confidentiality, as required by law, or because of a higher competing interest. Examples of legally required disclosures are public health hazards such as sexually transmissible diseases, which include HIV infections. The law not only permits such reporting, the law mandates it. Even if there is no regulation on point, a doctor may need to disclose sensitive information to named third parties if actual harm can be prevented through disclosure. Under the facts above, disclosure to the wife is necessary because repeated advice to the patient has gone unheeded and harm is imminent. The fact that the wife is also a patient simply adds to this ethical and legal duty to disclose. Of course the patient may sue the doctor for any breach of confidentiality but the doctor will likely win. There is a much greater risk for a lawsuit from the wife if the doctor fails to disclose and as a result, harm comes to the spouse. In such a lawsuit, the physician would likely lose.

ANSWER TO QUESTION 17: A, B, C

Choice A is correct because prescribing a drug without the proper indication amounts to breaching the standard of care, unless it is for an 'off-label' use, and there is scientific support for using the drug in that manner. The informed consent doctrine requires that physicians discuss all material risks, including rare but serious risks. Patients are assumed to have little or no knowledge of medications and they have no duty to inquire about side effects. The doctor, on the other hand, has an affirmative duty to warn of these side effects. In a malpractice case alleging lack of informed consent, the defense cannot plead contributory negligence. D is therefore incorrect.

The "Learned Intermediary" Doctrine stipulates that the doctor, not the pharmaceutical company, is liable for medication-related injuries as he/she is a learned professional who directly communicates with the patient and who does the actual prescribing. This puts the doctor in the hot seat for a drug reaction, although the drug company is frequently dragged into the lawsuit because it has the deep pocket. In some instances there is a separate duty of the pharmaceutical company to directly warn the consumer, e.g., immunization hazards.

ANSWER TO QUESTION 18: C, E

Strictly speaking, the doctor's negligence was a 'but-for' factual cause of the injuries. That is, without the doctor's negligent prescription in the first place, the plaintiff would not have suffered the harm. But Answer A is only partially correct, as the doctor will probably escape liability because the accident was an independent third force that intervened between his negligence and the injury, called a superseding cause. This requires the jury to deem the accident and outcome to be unforeseeable. A superseding cause will absolve the original tortfeasor (the doctor) from liability. In other words, Dr. DeSouza's negligence was not a proximate cause of Tony's injuries. C is therefore correct.

On the other hand, in the unlikely event that the accident was felt to be foreseeable, then it becomes a concurring cause and the doctor is liable (Choice D).

Of course the obvious party to sue is the careless driver. But he may be without insurance and unable to pay damages. The doctor, on the other hand, is likely to have substantial malpractice insurance coverage. Yes, the deep-pocket at work!

ANSWER TO QUESTION 19: NONE CORRECT

Informed consent is not the same as a consent form. The former is a process undertaken by the doctor to explain the benefits and risks of harm of a given procedure. It is required by law. The consent form, on the other hand, is merely documentation that such a process had taken place. A signed form where little or nothing was explained to the patient would not constitute informed consent. The patient does not have a duty to ask about risks, whereas the doctor always has an affirmative duty to inform.

Frightening material risks are sometimes withheld from overly anxious patients. This exception to disclosure of risks is called the therapeutic privilege but it should rarely be invoked. The facts must clearly indicate that disclosure would be harmful to the patient's overall well being. This case does not rise to such a level, and the doctor is unlikely to get away with this excuse.

The emergency exception does not apply here as there was time for obtaining consent.

Whether the physician-oriented standard (also called professional or medical standard) or the patient-oriented standard of disclosure will govern depends on the jurisdiction. Increasingly, states are replacing the medical or professional standard with the patient-oriented standard. Under this tough disclosure rule, what a reasonably prudent person in the patient's situation would have wanted to know will determine what the doctor needs to disclose. Hindsight is not a valid defense. Suits based on lack of informed consent are won where there is evidence of lack of risk disclosure, and where a reasonably prudent person in the patient's position would not have undergone the procedure had such risks been divulged.

ANSWER TO QUESTION 20: C, E

A verbal consent or even one that is implied, is legally valid, but proving that it was given is another matter. At trial, which is typically years down the road, the jurors may be reluctant to take the doctor's word. This is why a signed consent form is vastly preferred, because without documentation the plaintiff can argue that no risks were disclosed. Hence, in this case, entering a note in the chart would be good advice. It's too late to go back and get a signed consent, and backdating is of course a no-no.

Informed consent issues are usually raised in conjunction with alleged substandard treatment that resulted in an otherwise avoidable injury.

ANSWER TO QUESTION 21: E

None of the principles cited are relevant here. Therapeutic privilege is where disclosure of risks may prove detrimental to the patient's overall well being, so the doctor has the privilege to withhold such information. The gynecologist should in fact wait for another time to do the surgery even if it means that the patient will need to be re-anesthetized. The older medical or paternalistic ('doctor knows best') model has given way to the autonomy or self-determination model (patient has the last word). Answer A is attractive, but is no longer an acceptable choice. The 'best interest' approach is applicable only in an emergency situation where consent cannot be obtained in a timely manner. If a patient's wishes are unknown or unknowable, e.g., in a neonate, consent is still required from a legal surrogate decision-maker. Implied consent cannot be assumed unless the circumstances clearly so indicate, e.g., patient extending arm for venipuncture can be said to be giving implied consent for the procedure.

ANSWER TO QUESTION 22: A, D

The doctrine of informed consent requires the healthcare provider to inform patients about procedures, alternatives, and material risks. In order for patient consent to be meaningful, it has to be an informed one. Disclosure of material risks is an especially important part of informed consent, and this applies equally to surgical procedures as well as prescription medications. Invasive procedures or hazardous drugs warrant a more detailed explanation of the risks of harm. If it is shown that a reasonably prudent person would want to know about the risk of hypoglycemia and its effect on driving, and if the plaintiff can show that had she been warned, she would not have used insulin (or continued to drive), then she will likely prevail in a lawsuit against the doctor. Malpractice damages can be compensatory or punitive, and they can be based on physical injuries as well as emotional ones. Although the plaintiff in this case suffered no apparent physical injury, she can still claim damages for emotional distress, pain and suffering, etc.

The injured driver can of course sue her for negligent operation of a motor vehicle, but he may not prevail if the accident occurred through no fault of hers. This is why it is likely he will sue the doctor instead, the legal theory being that had the doctor warned her about hypoglycemia (and the precautions to take once symptoms begin to develop), she would not have struck his car. The fact that there is no doctor-patient relationship between him and the doctor is immaterial. The case will ultimately rest on whether the doctor owed a duty to him, a third party, independent of the doctor-patient relationship. Case law suggests that there may well be such a duty under the facts of this case and his lawsuit has a reasonable chance of success.

ANSWER TO QUESTION 23: B, C, D, E

This is a case of product liability. To be sure, the doctor did not prescribe it, but he did give his support, and the patient may have relied on his approval in taking the supplement. Just because it's an over-the-counter (OTC) herbal supplement does not altogether absolve the physician from blame since he was providing medical advice in his professional capacity when responding to the patient's question. This will become an increasingly frequent issue in malpractice litigation because of the widespread and growing use of alternative or complementary medicine. Totally unregulated, some of these OTC supplements are bound to result in harmful effects to the consumer. The injured party will naturally sue the manufacturer as well as the drugstore for putting the item into the stream of commerce.

Still, Ms. Holistica will have to prove that the preparation caused the injury, else the defendant will escape liability. Here the facts indicate hepatic failure, but it remains the plaintiff's burden to produce evidence that it was caused by the use of Slim-You.

ANSWER TO QUESTION 24: A, E

An emergency is an exception to the informed consent doctrine, but only if the next of kin cannot be readily found. Here we have a true emergency and time is of the essence. Performing surgery under the facts of this case is therefore permissible in the absence of consent. Forcing blood into this patient is a different matter. The facts are clear that he does not want blood because of religious beliefs. Denying him blood may lead to an otherwise preventable death but the courts have repeatedly ruled that this is a situation where patient autonomy trumps beneficence. Spousal disagreement is generally insufficient to override a patient's firmly held treatment preference.

ANSWER TO QUESTION 25: B, D

All medical entries should be made contemporaneously. A late or separate entry should be so identified. It is a bad idea to put off charting to another day, as it may be forgotten, and new events have a habit of overtaking the busy doctor. Charting is particularly critical when there is any hint of a malpractice complaint. Note that this case concerns a patient who may have developed a myocardial infarct that was 'missed' in the doctor's clinic. In his defense, the doctor did obtain a normal EKG (which may or may not be enough, depending on the clinical presentation and the patient's risk for a coronary event). If a suit is filed at a later date, the documentation of a normal EKG will serve as a good defense.

Medical records are crucial to defend a doctor from a malpractice claim. Without them, there is virtually no chance of escape. Treat your records as a true friend rather than a nuisance.

ANSWER TO QUESTION 26: C, D, E

Incompetence as applied to physician conduct is not a legal term of art. It may be the case that no nodule was palpable at the first visit, and whether or not the surgeon should have ordered a mammogram would depend on factors including family history, time of menstrual cycle and age. Importantly, we need to know whether the national guidelines recommend routine mammography in women under the age of 40. Asking the patient to return in three months for a recheck may represent a reasonable standard of care.

That the patient forgot her appointment may make her contributorily negligent, especially if the surgeon had reminded her just before or after the date (a good office practice).

The pathologist's error appears to constitute a substantial factor in her eventual disease spread. This puts the pathologist at serious risk for liability. Many jurisdictions would consider this loss of a chance sufficient proximate cause.

Finally, the doctrine of informed consent requires explanation of the procedure, alternatives including the option of non-treatment, and material risks — increasingly from a prudent person's viewpoint rather than from the doctor's viewpoint. Here, the patient is alleging that both the surgical goals and alternatives were not discussed. This makes a prima facie case for lack of informed consent.

ANSWER TO QUESTION 27: C, D

Breach of patient confidentiality is both a legal and ethical wrong. Privileged medical information cannot be disclosed except under well defined circumstances. In normal circumstances, unconsented disclosure of a patient's diagnoses to the employer is NOT permitted. B is incorrect.

However, patients may consent to the release of confidential information, or may waive confidentiality by putting their medical conditions at issue (records can then be viewed by experts, attorneys, etc.). Additionally, the law requires disclosure of privileged medical information where there is a substantial risk to an individual or the public, e.g., certain contagious diseases. In this case, alcohol abuse in a school-bus driver can reasonably be said to constitute a serious risk of harm to children, so disclosure may be preferable to confidentiality. The physician should first inform the patient that unless he abstains from further drinking and enters a rehabilitation program, he will have no choice but to report to the employer in the name of public safety. Although you have breached patient confidentiality by reporting to the employer, no legal liability will result from your action. Choice A is therefore incorrect.

If the patient works as an office clerk, then the doctor no longer has as strong a reason to report to the employer and he will likely lose the lawsuit if he were to breach confidentiality. The doctor should, however, emphasize to his patient the dangers of driving while intoxicated, and may consider reporting to the department of motor vehicles if the patient remains recalcitrant, especially if there is a past history of drunk driving.

For the usual sick-leave note, the doctor may simply state that the patient is to be excused from work 'for medical reasons' for a time period, without specifying the diagnosis. That's the usual case, but the facts here warrant a fuller disclosure of the medical conditions.

ANSWERS TO QUESTION 28: C, D, E

There are many legal risks in using the Internet in clinical medicine. Depending on the facts, a court may find that a doctor-patient relationship has been formed. Factors may include knowledge of names of subscribers, frequency of interactions, specificity of queries, and so on. In particular, a subscription fee is likely to be construed as evidence of soliciting and accepting a more committed interaction, so it places the operator of the website at greater legal risk. A specific disclaimer is a standard precaution but may not be enough to protect against a lawsuit.

Other than negligence, a viewer may also claim breach of confidentiality or privacy. E-mail not uncommonly ends up in the wrong mailboxes, and risks exposing sensitive medical information to strangers. Encryption of all identifiable electronic correspondence is therefore a good idea. Because e-mail preserves a record of what was asked, said, or recommended, they constitute powerful evidence supporting — or damning — the doctor.

ANSWER TO QUESTION 29: C, D

Retinal detachment is an ophthalmologic emergency and Dr. Moon should have promptly referred the patient to a specialist. His failure to arrange for a stat consult constitutes a breach of the standard of due care. As a GP, Dr. Moon may not have been expected to confirm or treat the condition but he has a duty to recognize the diagnosis and to make an immediate referral.

However, the patient must still show that were it not for the delay, she would have retained her vision. That is, she must show that the GP's negligence was a proximate cause of her blindness. She will need to produce expert testimony that the 36-hour delay adversely affected the outcome. The facts in this hypothetical question were taken from an actual malpractice case where the lower court found the GP to be negligent but the appellate court overturned the decision because causation was not proved.

ANSWER TO QUESTION 30: D

Consent is unnecessary in an emergency only if it cannot be readily obtained. This is not the case here, as the parents are available and are, in fact, refusing to give consent. The boy is too young to give consent, or even assent, i.e., to agree, to the procedure. Giving in to the parents would mean harm to the child who is too young and legally incompetent to protect himself. Answer D is therefore best under the circumstances, notwithstanding the risk that the parents might file a suit against the surgeon. If there is time, one would petition for a court hearing, which will most likely rule in favor of surgery. However, the medical facts suggest that immediate surgery is critical and a delay will place the child in peril of loss of the testicle. Unless an immediate court hearing can be arranged, option E is not a good choice.

ANSWER TO QUESTION 31: C

It is better to have a trusted friend who has a good attitude towards covering for your patients than a skilled but overworked colleague who considers it another burden to shoulder. Lawsuits begin with poor service, frustration, and anger, even before any substandard care is provided. The covering colleague can easily put the original doctor at risk. Good communication skills and proper attitudes are every bit as important as clinical knowledge.

It is generally insufficient to simply refer all calls to the emergency department. Doctors generally have a duty to secure coverage for their patients even when they are unavailable or not 'on call.' Of course patients should always be reminded that if they have a serious symptom or are unable to contact the doctor on call, they should go to the nearest emergency department for professional help.

ANSWER TO QUESTION 32: ALL CORRECT

Wishing to sue may begin even before any actual mal-occurrence has taken place. During the course of a medical encounter, especially in a hospital setting, there are numerous opportunities for the patient and family to feel uncared for, frustrated, and angry. These include medical bills, long waiting time, rudeness, inefficiency, confusion and so on. Then an unexpected adverse medical event occurs. How well do we respond? The doctor with poor bedside manners and who is arrogant and rushed is at increased risk for a lawsuit. Patients and families rarely sue a doctor they like, even though the outcome is bad.

Poor record-keeping makes malpractice defense that much harder. Remember the adage: 'Good records, good defense.'

ANSWERS TO QUESTION 33: IT ALL DEPENDS ON YOUR POLICY

All professional liability policies should cover negligence on the part of the doctor and employees, and peer review risks, although the hospital typically provides this latter coverage as well. However, do no take this for granted and be clear of the scope and exclusions of your policy. Most malpractice insurance policies do not cover premise liability (for accidents that occur on the premises that have nothing to do with medical care, e.g., tripping on the carpet or falling off a chair in the waiting-room). You, therefore, will need a separate premise insurance policy.

Malpractice insurance policies nowadays are usually claims-made policies. That is, when you retire and are no longer in practice, you will have to buy 'tail-coverage' to protect you from a malpractice lawsuit from a past event. Claims-made policies expire once you are no longer insured with the company.

Take a few minutes to understand the limitations of your professional liability policy. Make sure it's current and meets all of your needs. And never go bare. As retirement approaches, purchase tail coverage for the protection you deserve — just in case.

ANSWERS TO QUESTION 34: B, C

The Good Samaritan Law immunizes aid-givers for negligent acts, but there may be no immunity when such activities occur in a hospital setting. That is, doctors treating a stranger on hospital grounds may still be held to the same standard of care expected of a healthcare provider with a legal duty to treat (the usual provider-patient relationship). Nurse Nightingale and Dr. Goodnight may therefore be liable if their actions were shown to be negligent (not clear from the facts), and which resulted in harm (also not clear from the facts). The nurse working in a hospital setting obviously needs an update in CPR skills. Mr. Passerby, being a lay person will be held to a lower standard than the healthcare providers, and therefore will likely escape liability. Besides, breaking of ribs is not an uncommon result of vigorous chest compression.

Whether six minutes is too long a wait will be determined at trial by expert testimony. It appears to be a delayed response, but there may be extenuating circumstances, e.g., when was the Code-500 called, and was there another Code-500 in the hospital at the time?

ANSWERS TO QUESTION 35: ALL CORRECT

Choice A is correct as the law of malpractice will find the defendant liable only if injuries resulted from the negligent act.

Your office assistant should preferably be trained to recognize abnormal test results. At the minimum, there should be a system for the physician to see all reports before they are filed away, and then only after you have initialed the reports to indicate you have seen them. The assistant's liability will be imputed to you, the employer, under *respondeat superior*.

The lab owes an independent duty to inform the physician when there is a critically high or low value, and will be named as a co-defendant.

Overlooked reports account for many instances of patient harm, and these errors are preventable.

SECTION V

GLOSSARY OF LEGAL TERMS

Glossary of Legal Terms

Abandonment. The unethical and nonconsensual withdrawal of patient care by the treating doctor, i.e., unilateral severance of the doctor-patient relationship. Also used in property law to indicate relinquishment or surrender of property or rights attached.

Abrogation. The annulment or repeal of a law or legal doctrine, usually by legislation.

Abuse of Process. To make improper use of legal processes or proceedings, or to employ them in a manner contrary to the natural or legal rules for their use.

Advance Directives/Advance Medical Directives. Commonly known as 'Living Wills.' These are written documents designed to direct a patient's healthcare decisions, including end-of-life medical treatment, when he/she is no longer able to communicate. This term also encompasses healthcare powers of attorney.

Affidavit. A voluntary sworn written statement made by a person before a commissioner of oaths, a notary public, or someone with the appropriate authority.

Affirmative Defense. An assertion made by the defendant which constitutes a defense to an otherwise meritorious claim brought by the plaintiff.

Agency. Relationship in which one person (the agent) can legally act for, and legally bind another (the principal), with the latter's authority.

Allegation. Asserted but yet unproven fact, act or circumstance.

Alternative Dispute Resolution (ADR). Means of resolving disputes other than litigation. Examples include mediation and arbitration.

Alternative Medical Liability Act (AMLA). Bill introduced to the U.S. Congress in 1984 regarding a modified no-fault solution to medical malpractice claims.

Americans with Disabilities Act (ADA). Federal disability discrimination law protecting the rights of Americans with disabilities.

Antitrust Laws. Federal and state statutes that prohibit anti-competitive conduct such as price fixing and monopolies.

Appeal. A challenge of the decision of a lower court before a superior court, i.e., an appellate court.

Appellant. The one who files the appeal.

Appellee. The one against whom an appeal is taken.

Arbitration. A process of dispute resolution outside the court system wherein an impartial (third) party, the arbitrator, is accepted by both contending parties as the decision-maker and whose decision is final. Depending on the jurisdiction, the arbitrator's decision may or may not be subject to an appeal before a court of law.

Assault. An intentional act that causes apprehension of a harmful or offensive contact, e.g., shaking a fist under another's nose.

Assumption of Risk. An affirmative defense to a claim that bars the plaintiff from recovery for an injury when the plaintiff: (1) has knowledge of the dangerous nature of the activity or condition; (2) appreciates the nature or extent of the danger, and (3) voluntarily exposes himself or herself to the danger.

Battery. An unlawful intentionally harmful or offensive wounding or touching of another.

Borrowed Servant. See "Captain of the Ship" Doctrine.

Breach of Duty. In the law of torts, this refers to conduct of the tortfeasor (wrongdoer) falling below the standard of care expected of him/her under the circumstances.

Burden of Proof. The affirmative duty of proving a fact or facts in dispute. The legal burden of proving the claim and the loss suffered almost always falls on the plaintiff.

'But for' Test. A test of causation in tort. The plaintiff can establish causation if he/she can prove that he or she would not have suffered the injury complained of, but for the conduct of the defendant.

'Captain of the Ship' Doctrine. Generally refers to the situation where a surgeon is in charge of — and therefore responsible for — the actions of his surgical assistants who may not be his employees but employees of the hospital ('borrowed servants').

Case Law. Adjudicated cases that become legal precedents which the court may follow in subsequent cases. Laws other than those established via the legislative process (called statutes/legislation).

Causation. The factual link between the acts of the tortfeasor and the injury.

Cause in Fact (Factual Cause). The act (or omission) without which the injury would not have occurred.

Cause of Action. An articulable right to judicial relief based on facts that constitute a cognizable civil wrong. In lay terms, a claim.

Caveat Emptor. Latin maxim meaning 'Let the buyer beware.'

Charitable Immunity. Outdated doctrine that holds a charitable organization such as a hospital immune from lawsuit.

Civil Action. A legal claim for monetary compensation, as opposed to a criminal action brought by the State (prosecutor) against a defendant that may subject the accused to a prison term.

Claims-Made Insurance Policy. Covers claims made during the insured period but not if the policy is no longer in effect (See 'Occurrence Policy' and 'Tail Coverage').

Class Action. A lawsuit filed by one person on behalf of himself or herself and all others who are similarly situated.

Clear and Convincing Evidence. Level of proof that is higher than a preponderance but less than beyond reasonable doubt. Used in immigration cases, denial of visitation rights, etc. In some states, this level of proof is required before a surrogate decision-maker is allowed to discontinue life-sustaining treatment.

Collateral Source Rule. A rule which excludes as irrelevant any evidence of a plaintiff's alternative or additional sources of payment for losses or damages and which allows an injured person to receive compensation from a source wholly independent of the tortfeasor, even if damages are otherwise collected from the tortfeasor.

Common Law. The body of principles and rules derived from the judgments and decrees of the courts, particularly the unwritten law of England. Today, it is often used to refer to law that is not statute-based, i.e., case law.

Comparative Negligence. Theory of law wherein a plaintiff's damages are reduced proportionately to the plaintiff's degree of fault in causing the injury, rather than completely barring recovery.

Compensatory Damages. Monetary awards that compensate or restore the loss caused by the wrong or injury, i.e., damages to put the victim in a position as if the wrong or injury had never occurred, and make the plaintiff 'whole'.

Complaint. The initial pleading that states the cause of action and which commences a lawsuit.

Confidentiality. Ethical and legal requirement in a relationship such as the doctor-patient relationship wherein the secrets of the patient are entrusted to the healthcare provider to be kept confidential and not for public disclosure.

Consent. Agreeing to or giving permission.

Contempt of Court. An act of disobedience of the rules, orders, or process of a court that diminishes the dignity and authority of the court in its administration of justice, e.g., not carrying out an order of the court.

Contingency Fee. Arrangement between attorney and client wherein attorney agrees to represent the client on the basis that the amount of his fees will depend on the outcome of the litigation. If no damages are awarded, no fee is assessed.

Contributory Negligence. Conduct by a plaintiff which is proven to be a contributory cause of the injury. Such negligence may serve to free the defendant from liability.

Coroner's Inquiry. An inquisition or examination into the causes and circumstances of any death happening by violence or under suspicious/unusual circumstances, held by the coroner in the coroner's court.

Counterclaim. The defendant's claim against the plaintiff usually seeking to reduce or extinguish the plaintiff's claim.

Cross-examination. The questioning of a witness by opposing counsel at trial or upon taking a deposition, to advance the case of the cross-examiner's client or to attack the case of his client's opponent.

Damages. Monetary compensation awarded by the court to a person who has suffered loss or injury through the unlawful act or omission, e.g., through the negligence, of another.

Defamation. Communication calculated to injure reputation or to expose a person to ridicule, scorn or contempt; to harm a person's reputation in the eyes of the public or of a particular audience. Includes both libel (written) and slander (verbal).

Defense. Facts or arguments that support the position offered by the defending party in answer to plaintiff's claim so as to deny him/her any legal relief or remedy.

Deposition. The testimony of a witness taken before trial by lawyers from both sides which is preserved *verbatim* for subsequent use at trial.

Directed Verdict. Instruction by judge directing the jury to decide in favor of the party moving for such a verdict because of overwhelming evidence.

Discovery. The disclosure of everything that is relevant to the claim and the defense. This pre-trial process seeks to ascertain, catalog and disclose all documents and information which might have been previously unknown to either party, so as to avoid 'trial by ambush.'

Due Care. Sufficient care within the given circumstances so as to rise above substandard or negligent conduct.

Durable Power of Attorney. A power of attorney is an instrument authorizing another to act as one's agent. The agent's power is revoked upon the death of the principal. A durable power of attorney accords legal power which springs into effect upon the principal's incapacity and which is not extinguished by death.

Duty. A legal or moral obligation.

Duty of Care. A legal obligation to take care in one's dealings with another.

'Eggshell Skull' Rule. A tortfeasor is liable for all harm done to the victim even if the victim has a predisposing condition and suffers more serious injuries than a normal person.

Employment Retirement Income Security Act of 1997 (ERISA). Extensive federal statutes that regulate pension plans and employee benefits, including healthcare benefits.

Evidence. An item of proof such as the testimony of a witness, the contents of a document, an object, or other direct or indirect proof introduced at trial in support of the legal position taken by a party.

Expert Witness. A witness who possesses specialized knowledge of a skill, science, or art that goes beyond the knowledge held by ordinary individuals.

Fiduciary. A trusted person (e.g., a trustee or an agent) who is expected to conduct himself or herself in a scrupulous and fair manner in his relationships with the client, watching out for and protecting that person's best interests.

General Damages. Compensation for loss flowing naturally from the wrongful act, e.g., compensation for pain and suffering or the loss of amenities.

Good Faith. An honest belief, the absence of ill will, and the absence of design to defraud or gain an advantage.

'Good Samaritan' Doctrine. Theory of law recognizing that one who comes to the aid of a person in imminent and serious peril is not liable for negligence so long as the attempt is not reckless and there is no expectation of monetary reward.

Gross Negligence. Conduct worse than simple negligence where there is reckless disregard for the consequences.

Guardian. Court-appointed legal representative of a person who is declared incompetent.

Hearsay. An out-of-court statement offered at trial by someone other than the declarant to prove the truth of the matter asserted. For example, where the witness testifies that something is true because he/she heard somebody else say something to that effect.

Holding. A ruling of the court.

Incompetence. Term used to designate loss of legal capacity to manage one's personal and business affairs, including such things as making contracts or medical decisions.

Indemnity. Insurance term that denotes an obligation to pay or reimburse for losses.

Independent Contractor. One who is contractually obligated to perform a task, but who is in general control of his or her own methods and means of doing so rather than abiding by the controlling rules of an employer.

Informed Consent. Consent given by a patient to a doctor, e.g., for surgery, based upon a full disclosure of facts, including the diagnosis, proposed treatment and material risks that are needed to make the decision freely and intelligently.

Injunction. A legal remedy granted by a court that compels a party to perform a specific act (a mandatory injunction) or restrains a party from performing a specific act (a prohibitive injunction).

Intentional Tort. An intentional civil wrong that is perpetrated by one who intends to commit the wrong. This contrasts with negligence where the tortfeasor may not intend to commit the wrong but merely fails to exercise the required standard of care.

Interrogatories. Part of the discovery process where questions are formally asked and answered by the parties in preparation for trial.

Joint and Several Liability. Doctrine that holds all joint tortfeasors individually responsible for paying the entire amount of loss incurred by the plaintiff, but the individual may in turn sue the other tortfeasors for contribution.

Judgment. Official decision of the court.

Judgment Notwithstanding the Verdict (JNOV). Overturn of the verdict by the trial judge upon motion of the moving party because the weight of the evidence failed to support jury's verdict. Similar to a directed verdict, which is rendered before trial is completed.

Jurisprudence. The science or philosophy of law.

Jury. Men and women, called jurors, selected according to law and sworn to determine the evidentiary truths presented at trial.

Last Clear Chance. A principle applied in automobile law through which a plaintiff may recover from a defendant motorist for injuries or damages suffered, notwithstanding his or her own contributory negligence, if the motorist had the last opportunity to avoid the accident through the exercise of reasonable care. The doctrine is largely abandoned in the law of medical negligence.

'Learned Intermediary' Doctrine. Legal doctrine wherein the treating doctor remains liable for injuries arising out of the prescription of medical products, but where manufacturers (such as the pharmaceutical companies) are not.

Legal Cause. A calculus regarding the foreseeabilty of results or remoteness of damage, so as to assign fault or responsibility. Also referred to as proximate cause.

Libel. Written form of defamation.

Litigant. A party to a lawsuit.

Litigation. A lawsuit — legal action for the purpose of enforcing a right or seeking a remedy.

Locality Rule. The standard of care that is adhered to within the locale of medical practice rather than a national standard of care.

Loss of Consortium. Recoverable non-economic damages compensating a spouse for loss of conjugal relations resulting from plaintiff's injuries.

Malfeasance. An unlawful or wrongful act.

Malice. Purposeful wrongful act with apparent intent to injure.

Malicious Prosecution. Civil or criminal action, often begun in malice, and taken without probable cause that the accusations can be proven.

Malpractice. Negligent or otherwise improper conduct, i.e., unreasonable lack of care in discharging duties owed to a patient or client, by a professional such as a doctor, lawyer, or accountant.

Master-Servant Doctrine. In modern day practice, this refers to the employer-employee relationship where the employer (master) has the right to select the employee (servant), and the power to control and direct his/her job duties.

Mediation. Method of dispute resolution where a third party mediates between the contending parties towards persuading them to settle their dispute. The mediator himself or herself does not pass judgment or render any decision.

Medical Negligence. Synonymous with medical malpractice.

Mitigate. To abate or reduce, e.g., a penalty, injury or damages.

Motion. A formal request made by an attorney for an order or ruling by the judge.

Negligence. An act or an omission, which falls below the standard established by law or is otherwise expected from a reasonably prudent person under the given circumstances.

Negligent Infliction of Emotional Distress. A cause of action filed by a person who suffers severe emotional distress, e.g., psychiatric injury, as a result of witnessing a tortious act committed against another person — usually a closely related victim.

No-Fault. A type of insurance best exemplified by automobile insurance, in which each person's own insurance company pays for injury or damage up to a certain limit regardless of who was at fault.

Non-Delegable Duty. A legal duty that cannot be delegated to someone else.

Nonfeasance. Failure to act where duty requires action.

Nose Coverage. When switching from one carrier to another, the insured buys nose coverage to cover claims that may be filed after a previous claims-made policy has lapsed but before new coverage becomes effective.

Occurrence Insurance Policy. Covers claims occurring during the period covered by the policy, irrespective of when these claims are filed, even if the policy is no longer in effect.

Ostensible Agency. An implied agency where the person is held out to be an agent though not explicitly designated as such.

Pain and Suffering. Injuries that comprise physical discomfort and emotional distress, e.g., psychiatric injury, which are recoverable as elements of damages in torts.

Patient-Physician Privilege. A legal right of a patient not to have his medical secrets divulged by the treating doctor at trial.

Penal Code. Collection of criminal (penal) laws of the state or country.

Perjury. Intentionally false testimony given under oath and punishable by law.

Plaintiff. The person who files a civil lawsuit to seek relief for injuries suffered.

Power of Attorney. Authority of one to act as the legal agent or representative for another. The agent, who does not have to be a lawyer, is said to have Power of Attorney.

Preponderance of Evidence. Evidence which, on the whole, is more convincing than that offered by the opposition. Synonymous with 'more probable than not' and 'more likely than not.'

***Prima Facie* Case.** The minimum amount of proof required to establish the existence of a cause of action.

***Prima Facie* Evidence.** Evidence sufficient to establish a given fact and which if not rebutted or contradicted will remain sufficient.

Privilege. A legal shield that confers protection on one who holds the privilege. For example, the doctor-patient privilege is a rule of evidence that protects the patient, the holder of the privilege, from having his medical history revealed by the doctor.

Probable Cause. Sufficient evidence to warrant a search or an arrest in a criminal case.

Products Liability. Legal causes of action based on injuries arising out of the use of a defective product.

Proximate Cause. An act or omission that produces a result without which the injury would not have occurred, providing that the natural sequence of the injury is unbroken by an intervening cause. Also called legal cause.

Punitive or Exemplary Damages. Damages that are awarded to the plaintiff over and above compensatory damages for aggravated or malicious acts. The purpose of such damages is to punish the tortfeasor for the egregious conduct.

Quantum. An amount. In litgation, commonly used to refer specifically to the amount of damages.

Reasonable Person. A legal term used to objectively describe an ordinarily prudent individual whose conduct under the circumstances is taken as denoting the standard.

Res Ipsa Loquitur. *'The thing speaks for itself.'* A means of proving the negligence of the defendant if the circumstances in which the injury was caused was in the defendant's exclusive control.

Respondeat Superior. *'Let the master answer.'* Maxim that an employer (master) is liable for the wrongful acts of his employee (servant) if the negligent act was carried out by the employee within the scope of employment.

Restatement of Torts. A series of volumes authored by the American Law Institute that describes the law of torts, past and present. An important secondary source of legal authority.

Rule-11. A federal procedural rule of court that requires attorneys to 'reasonably inquire' into the plaintiff's claim before filing suit.

Rule of Avoidable Consequences. The victim is legally required to mitigate his/her injuries to avoid further aggravation of the condition.

Sanctions. Penalty affixed by law for a violation. Used as a verb, it means an assent or a reprimand.

Settlement. Mutual agreement by the parties to terminate the lawsuit usually in exchange for a sum of money and disavowal of fault.

Sherman Act. An act of the U.S. Congress that promulgates various laws against monopolies, price fixing, and other anti-competitive practices in commerce.

Slander. The oral form of defamation.

Special Damages. Associated with economic or pecuniary damages. Provides compensation for lost wages, medical expenses and other out-of-pocket expenses.

Standard of Care. The degree of care which a reasonably prudent person in a particular profession or position should exercise under same or similar circumstances. A fundamental element in the tort of negligence. A measure of whether the defendant has breached his legal duty of care.

Stare Decisis. Legal doctrine of abiding by previously decided cases (case precedent).

Statute. A written law enacted by the legislature of the state or country.

Statute of Limitations. A statute prescribing the time limitations within which a lawsuit must be brought or else it will be barred. Also referred to as the Limitation of Actions.

Stipulation. A party's agreement not to dispute a particular fact or circumstance.

Strict Liability. Liability that does not depend on actual negligence or an intent to harm, but is based on the breach of an absolute duty to make something safe. Applies especially in product liability cases in which a manufacturer, distributor, or seller may be liable for defective or hazardous products which are unreasonably risky.

Structured Settlement. A negotiated schedule of payment that usually involves periodic installments instead of a single lump-sum payment.

Subpoena. A written instrument issued by the court and directed to a witness requiring that person to give testimony in court or produce documents (*Subpoena Duces Tecum*) at a certain time and place.

Summary Judgment. Decision by the court without requiring the case to proceed to trial where there is no genuine issue of material fact and the moving party is entitled to judgment on his/her claim as a matter of law.

Summons. A written instrument issued by the court to the defendant notifying him/her that a legal action has been commenced by the plaintiff and calling for the defendant to state whether he/she wishes to oppose the claim.

Superseding Cause. An event that occurs between the defendant's act and the plaintiff's injury that breaks the chain of causation so that the defendant is no longer liable to the plaintiff. The law does not hold the defendant liable when an unforeseeable intermediary factor intervenes and leads to an unforeseeable injury.

Tail Coverage. That difference in insurance coverage between occurrence coverage and claims-made coverage.

Testimony. Evidence given by a competent witness under oath or affirmation, as distinguished from evidence derived from writings and other sources.

Therapeutic Privilege. The doctrine that a physician may withhold disclosure of information regarding medical risks if such disclosure will be detrimental to the patient's total care and best interest.

Tort. A civil injury or wrong affecting private citizens that is not based upon a breach of contract.

Tortfeasor. A wrongdoer in tort; the person who commits a tort.

Trial. A judicial examination and determination of issues between parties in accordance with the law of the land. May be a jury or non-jury (judge) trial.

Vicarious Liability. Indirect legal fault that is imputed from an employer-employee relationship or a principal-agent relationship.

Wanton. Describes conduct that is grossly negligent, with reckless disregard for the consequences, or malicious.

Warranty. A term in contract law equivalent to a guarantee.

Index

abandonment 33, 34, 321
abuse of process 83, 84, 87, 321
accreditation 8, 161, 218, 262
advance directive 112, 115–117, 321
affidavit 270, 321
affirmative defenses 23, 73, 74, 77, 78, 321, 322
agency 132, 150, 151, 210, 214, 247, 263, 286, 304, 321, 330
Alternative Dispute Resolution (ADR) 265, 321
Alternative Medical Liability Act (AMLA) 266, 275, 322
Alternative Medicine 131–133, 138
American Medical Association (AMA) 5, 24, 31, 41, 107, 127, 246, 266, 278
Americans with Disabilities Act (ADA) 322
anger 8, 12, 212, 213, 216, 222, 223, 225, 231, 247, 255, 315
antitrust 85, 322
appeal 6, 16, 21, 26, 44, 50–52, 65, 67, 84, 96, 99, 101, 105, 116, 120, 125, 128, 131, 135, 141, 145, 154, 158, 164, 241, 276, 278, 322
arbitration 18, 193, 270, 279, 321, 322
assault 21, 26, 88, 89, 99, 179, 195, 240, 288, 306, 322
assumption of risk 74, 76–78, 135, 141, 194, 285, 303, 322
aviation industry 217
awards 5, 15, 16, 58, 60, 161, 187–191, 269, 278, 324

bad faith 38, 66, 200
battery 21, 88, 89, 96, 99, 102, 103, 160, 195, 240, 288, 292, 306, 322
bedpan mutuals 15, 242
burden of proof 23, 49, 122, 303, 322
'but-for' 48, 49, 55, 269, 308

California 25, 30, 31, 33, 34, 41, 44, 48, 49, 52, 58, 59, 69, 73, 77, 80, 81, 92, 100, 101, 105, 106, 108, 115, 116, 119, 122, 126, 128, 140, 141, 144, 153, 164, 179, 189, 196, 207, 220, 237, 242, 247, 266, 271–273
capacity 88, 94, 112, 117, 162, 311, 327
caps 10, 262, 266, 272, 280
captain of the ship 151, 162, 322, 323
Cause in Fact (Factual Cause) 323
class-action 10, 139, 142, 323
code of ethics 25, 176
collateral source rule 269, 324
common law 21, 73, 75, 275, 324
community standard 24, 39, 42, 45, 289, 291
comparative negligence 54, 74, 75, 78, 303, 324
compensable event 274, 277, 278
compensatory damages 57–59, 62, 84, 141, 324, 331
competence 24, 131, 132, 198, 220, 261
complaint 66, 73, 82, 122, 152, 196, 202, 207, 210, 211, 213, 246–249, 251, 299, 312, 324

335

computerized medical records 172, 174, 264
confidentiality 33, 108, 112, 160, 171, 176–183, 186, 202, 213, 290, 295, 307, 308, 313, 314, 324
conflict of interests 152, 185
consult 29, 120, 172, 183, 210, 295, 296, 314
contempt of court 84, 180, 324
contingency fee 6, 9, 10, 18, 79, 210, 241, 266–268, 271, 324
contract 6, 18, 21, 27, 140, 149, 151, 178, 270, 276, 279, 327, 333
contributory negligence 23, 53, 54, 75, 76, 78, 97, 164, 300, 303, 308, 324, 328
corporate liability 148, 156, 277
counterclaim 83, 249, 325
countersuit 79–81, 83–85, 87
criminal 22, 25–27, 42, 79, 123, 127–129, 165, 172, 178, 183, 217, 240, 254, 323, 328, 330
cross-examination 254, 255, 325

deep pocket 151, 241, 268, 308, 309
defamation 83, 84, 179, 180, 325, 328, 332
defensive medicine 11, 200
dentist 22, 34, 39, 61, 89, 159
deposition 64, 83, 246, 250, 252, 253, 255, 257, 325
deterrence 10, 264, 265, 277
disclosure 19, 77, 90, 91, 93, 95, 96, 98–100, 102, 103, 136, 137, 161, 172, 175–180, 184, 217, 221, 222, 224–226, 228, 230–235, 291, 308–311, 313, 314, 324, 325, 327, 333
discovery 19, 67, 74, 80, 180, 219, 227, 240, 249–253, 262, 325, 327
dispute resolution 270, 322, 329
doctor-patient relationship 9, 17, 22, 28, 29, 31–33, 91, 112, 145, 152, 159, 160, 163, 165, 176, 182–184, 186, 201, 212, 215, 221, 223, 247,

273, 277, 284, 285, 287, 293, 295, 301, 302, 304, 311, 314, 321, 324
documentation 93, 109, 153, 167, 200, 209, 309, 310, 312
due care 37, 63, 136, 144, 163, 183, 184, 186, 274, 295, 314, 325
durable power of attorney 94, 112, 325

economic loss 276, 277, 279
eggshell skull rule 53–55, 326
Emergency Medical Treatment and Active Labor Act (EMTALA) 32, 120
empathy 201, 207, 208, 226, 229, 231, 261
Employee Retirement and Income Security Act (ERISA) 153
eRisk 183–185
ethics 25, 31, 32, 65, 109, 111, 113, 115, 118–121, 125, 176, 262, 306
exemplary damages 59, 82, 331
expert testimony 22, 34, 38, 39, 41–43, 45–47, 52, 63, 65–70, 72, 99, 122, 231, 250, 287, 301–303, 307, 315, 317
expert witness 6, 39, 45, 47, 64, 66–70, 81, 82, 85, 86, 270, 326

factual cause 48–50, 55, 308, 323
failure to diagnose 52, 153, 188, 194, 196, 197, 300
fiduciary 92, 154, 326
Florida 4, 5, 26, 51, 66, 108, 115–117, 150, 191, 193, 269, 272
Food and Drug Administration (FDA) 106, 132, 146
frequency of claims 189, 239
futility 118–121, 130

gag rule 152
gatekeeper 69, 152
General Accounting Office (GAO) 13, 188
general damages 62, 122, 326
good faith 30, 31, 36, 79, 109, 200, 326

good samaritan doctrine 287, 326
gross negligence 25, 30, 31, 59, 136, 165, 304, 306, 326
guidelines 25, 42, 64, 66, 70, 107, 108, 118, 242, 262, 313

Hawaii 13, 24, 30, 31, 33, 41, 60, 70, 90, 92, 93, 96, 99–101, 104, 105, 108, 112, 119, 123, 127, 132, 134, 160, 174–176, 208, 237, 238, 242, 267, 271
Health Insurance Portability and Accountability Act (HIPAA) 172
Health Maintenance Organizations (HMOs) 152
Hearsay 327
Hippocratic Oath 177, 256
HIV/Aids 46, 92, 144, 193, 290, 308
hospital liability 149

implied consent 92, 112, 177, 292, 310
indemnity 11, 132, 242, 327
independent contractor 149–151, 156, 284, 286, 302, 304, 305, 327
informed refusal 88
injunction 327
Institute of Medicine (IOM) 8, 218, 264
institutional liability 148
intentional tort 21, 89, 240, 276, 288, 292, 327
Internet 7, 29, 146, 182–186, 288, 314

joint and several liability 60, 266, 268, 269, 327
Joint Commission for the Accreditation of Healthcare Organizations (JCAHO) 8

last clear chance 328
learned intermediary doctrine 145, 288, 305, 308, 328
legal cause 21, 48, 50, 51, 288, 301, 328, 331
litigation stress syndrome 11, 12
living will 94, 112, 321

locality rule 38, 328
loss of a chance 52, 55, 301, 313
lump sum 57–59, 91

malicious prosecution 79–82, 84, 87, 328
Managed Care Organization (MCOs) 152, 156
master-servant doctrine 329
mediation 270, 278, 321, 329
medical errors 8, 19, 23
Medical Injury Compensation Reform Act (MICRA) 59, 271
medical records 33, 34, 81, 83, 171–177, 180, 181, 183, 209, 210, 213, 247, 249, 250, 252, 255, 294, 312
minority rule 39
minors 74, 94, 174, 270
misconduct 24, 30, 66, 81, 133
mistake 12, 23, 125, 161, 214, 218–222, 225, 227–231

National Practitioner Data Bank 5, 12, 157, 242, 253
negligent infliction of emotional distress 56, 286, 301, 329
New York 5, 9, 10, 14–19, 25, 28, 29, 68, 84, 88, 89, 92, 119, 123, 124, 136, 142, 165, 166, 202, 222, 224, 264, 272
no-fault 6, 18, 261, 265, 266, 273–276, 278, 280, 322, 329
non-delegable duty 149, 156, 329
non-economic damages 18, 58, 59, 122, 266, 267, 272, 328

occurrence insurance 330
online 8, 184, 185, 218
Oregon 5, 126, 127, 130
ostensible agency 150, 152

pain and suffering 58, 59, 62, 122, 266, 273, 311, 326, 330
patient-physician privilege 330

patient safety 166, 173, 216, 218, 220, 233, 261, 264
peer review 65, 69, 70, 177, 180, 198, 240, 244, 277, 279, 297, 316
practice guidelines 42
pre-existing 53, 57, 273
premiums 3-6, 8, 11, 13, 15-17, 19, 20, 236-239, 243, 264, 269, 272, 275, 278
preponderance of evidence 22, 48, 254, 304, 330
prescription 40, 104, 121, 126, 136, 139, 144, 145, 173, 174, 181-183, 185, 206, 213, 287, 290, 291, 301, 305, 308, 311, 328
prima facie evidence 43, 330
privacy 84, 113, 160, 172, 173, 177-179, 182, 183, 186, 215, 295, 314
privileges 12, 18, 85, 109, 157, 192, 193
procedure 11, 16, 40, 57, 68, 81, 82, 85, 89-98, 101, 103, 112, 114, 149, 154, 162, 165, 172, 188, 192, 195, 200-202, 209, 220, 225, 227, 239, 244, 285, 291, 293, 298, 309-311, 313, 315
products liability 139-142, 144, 147, 149, 182, 272, 331
professional misconduct 24, 27
proof 22, 41, 100, 220, 254, 276, 301, 323, 326, 330
proximate cause 16, 23, 48, 50, 55, 60, 105, 129, 141, 164, 288, 301, 309, 313, 315, 328, 331
punitive damages 57, 59, 62, 82, 84, 122, 141, 144, 154, 193, 195, 272

quality 6, 9, 10, 30, 132, 140, 146, 152, 171, 185, 198, 209, 212, 214, 216, 218, 243, 261, 274, 277, 301

reasonable person 28, 37, 99-102, 105, 112, 283, 300, 331

reform 16-19, 58, 59, 166, 261, 266, 267, 269, 271, 272, 278, 280
remoteness 328
res ipsa loquitur 42, 43, 47, 49, 63, 194, 286, 331
respondeat superior 150, 151, 156, 163, 179, 194, 302, 317, 331
restatement of torts 77, 331
risk management 136, 167, 168, 169, 191, 221, 222, 279, 297
rule of avoidable consequences 54, 56, 331
rule 11 84, 85, 87, 331

safety 8, 45, 95, 106-109, 119, 132, 136, 144, 176, 179, 188, 209, 217, 218, 220, 233, 293, 295, 313
sanction 67, 84, 85, 87, 127, 178, 331
scientific evidence 68, 69
screening panel 270, 271, 278
settlement 5, 6, 9, 12, 57, 83, 167, 187-193, 195, 196, 240, 241, 246, 253, 268, 272, 276, 331, 332
severity of claims 7, 187, 188, 239
Sherman Act 332
special damages 62, 332
staff privileges 268
stare decisis 21, 332
statute 30-32, 36, 42, 44, 45, 57, 74, 77, 78, 82, 90, 95, 119, 125, 127, 132, 153, 160, 172, 175, 176, 179, 181, 266, 269-271, 283, 297, 298, 300, 304, 322, 326, 332
statute of limitations 19, 73-75, 78, 89, 332
strict liability 140, 141, 143, 144, 147, 149, 288, 305, 332
structured settlement 332
subpoena 180, 332
substandard care 20, 38, 48, 134, 149, 264, 277, 292, 315
summary judgment 75, 332

Supreme Court 16, 23–25, 32, 33, 45,
48, 49, 65, 68–70, 74, 76, 92, 96,
100, 101, 105, 113–116, 123–125,
127, 140, 141, 144, 154, 155,
272

tail coverage 161, 236, 240, 243, 244,
298, 316, 323, 333
telephone 29, 171, 184, 202, 215
testimony 42, 63–72, 85, 89, 113, 251,
254, 278, 330, 332
therapeutic privilege 95, 96, 99, 100,
292, 309, 310, 333
third party 32, 33, 36, 89, 105, 110,
176, 179, 249, 300, 301, 308, 311,
329
tort reform 17, 18, 21, 58, 60, 261,
266, 267, 270, 273, 278–280
tortfeasor 37, 51, 55, 59, 60, 62, 75,
76, 148, 150, 268, 269, 278, 301,
309, 322–324, 326, 327, 331, 333

trust 9, 13, 14, 20, 34, 112, 152, 161,
176, 207, 221–223, 225, 232, 233,
235, 255, 265, 307

vaccination 139
vasectomy 57, 100
verdict 5, 6, 9, 11, 16, 19, 39, 58, 61,
64, 84, 105, 120–122, 141, 142,
144, 145, 153, 189–196, 209, 210,
212, 241, 254, 265, 266, 325, 327,
328
vicarious liability 148, 150, 151, 156,
161, 162, 164, 194, 302, 305, 333
violation of statute 42, 44, 45, 47

waiver 43, 97, 108
warranty 133, 139, 140, 143, 144, 147,
333
wrongful birth 57
wrongful death 19, 40, 56, 80, 276
wrongful life 57